Lecture Notes in Computer Science   10990

Commenced Publication in 1973
Founding and Former Series Editors:
Gerhard Goos, Juris Hartmanis, and Jan van Leeuwen

Editorial Board

Marina L. Gavrilova · C. J. Kenneth Tan (Eds.)

Transactions on Computational Science XXXIII

Editors-in-Chief
Marina L. Gavrilova
University of Calgary
Calgary, AB
Canada

C. J. Kenneth Tan
Sardina Systems OÜ
Tallinn
Estonia

ISSN 0302-9743 ISSN 1611-3349 (electronic)
Lecture Notes in Computer Science
ISSN 1866-4733 ISSN 1866-4741 (electronic)
Transactions on Computational Science
ISBN 978-3-662-58038-7 ISBN 978-3-662-58039-4 (eBook)
https://doi.org/10.1007/978-3-662-58039-4

Library of Congress Control Number: 2018936192

This Springer imprint is published by the registered company Springer-Verlag GmbH, DE
part of Springer Nature.
The registered company address is: Heidelberger Platz 3, 14197 Berlin, Germany

LNCS Transactions on Computational Science

Computational science, an emerging and increasingly vital field, is now widely recognized as an integral part of scientific and technical investigations, affecting researchers and practitioners in areas ranging from aerospace and automotive research to biochemistry, electronics, geosciences, mathematics, and physics. Computer systems research and the exploitation of applied research naturally complement each other. The increased complexity of many challenges in computational science demands the use of supercomputing, parallel processing, sophisticated algorithms, and advanced system software and architecture. It is therefore invaluable to have input by systems research experts in applied computational science research.

Transactions on Computational Science focuses on original high-quality research in the realm of computational science in parallel and distributed environments, also encompassing the underlying theoretical foundations and the applications of large-scale computation.

The journal offers practitioners and researchers the opportunity to share computational techniques and solutions in this area, to identify new issues, and to shape future directions for research, and it enables industrial users to apply leading-edge, large-scale, high-performance computational methods.

In addition to addressing various research and application issues, the journal aims to present material that is validated – crucial to the application and advancement of the research conducted in academic and industrial settings. In this spirit, the journal focuses on publications that present results and computational techniques that are verifiable.

Scope

The scope of the journal includes, but is not limited to, the following computational methods and applications:

- Aeronautics and Aerospace
- Astrophysics
- Big Data Analytics
- Bioinformatics
- Biometric Technologies
- Climate and Weather Modeling
- Communication and Data Networks
- Compilers and Operating Systems
- Computer Graphics
- Computational Biology
- Computational Chemistry
- Computational Finance and Econometrics

- Computational Fluid Dynamics
- Computational Geometry
- Computational Number Theory
- Data Representation and Storage
- Data Mining and Data Warehousing
- Information and Online Security
- Grid Computing
- Hardware/Software Co-design
- High-Performance Computing
- Image and Video Processing
- Information Systems
- Information Retrieval
- Modeling and Simulations
- Mobile Computing
- Numerical and Scientific Computing
- Parallel and Distributed Computing
- Robotics and Navigation
- Supercomputing
- System-on-Chip Design and Engineering
- Virtual Reality and Cyberworlds
- Visualization

Editorial

The *Transactions on Computational Science* journal is published as part of the Springer series *Lecture Notes in Computer Science*, and is devoted to a range of computational science issues, from theoretical aspects to application-dependent studies and the validation of emerging technologies.

The journal focuses on original high-quality research in the realm of computational science in parallel and distributed environments, encompassing the theoretical foundations and the applications of large-scale computations and massive data processing. Practitioners and researchers share computational techniques and solutions in the area, identify new issues, and shape future directions for research, as well as enable industrial users to apply the presented techniques.

The current issue is devoted to research on computational geometry and computability, with applications in the Internet of Things, Bioinformatics, and Wireless Body Area Networks. It is comprised of seven papers, three of which were invited following the 18th International Workshop on Computational Geometry and Security Applications, CGSA 2017, held in Trieste, Italy, in June 2017.

The first four papers are devoted to finding computationally efficient solutions to open problems in the fields of Bioinformatics, the Internet of Things (IoT), and Wireless Body Area Networks (WBAN). The first article on IoT defines a world-wide cyber-physical system with a plethora of applications in the fields of demotic, e-health, goods monitoring, and logistics, and presents the use of cross-layer communication schemes. The second article introduces the Wireless Body Area Network (WBAN) as an upcoming research area where wireless sensors are implanted within a human body or attached to the body, with emphasis on making such networks energy efficient and stable. The third paper addresses the critical issue in the rapidly growing field of Bioinformatics of how to regulate and manage the enormous amount of new data to facilitate seamless access to vast biological information, through introducing a new gene classification method in DNA sequences. The forth paper addresses another classification problem, namely how to perform an automated classification of web blocks through a combination of machine learning and a model-driven feature extraction method.

The remaining three papers are extended versions of the conference papers, invited following CGSA 2017. They present research on sequential and parallelizable computations and discussion of Amdahl's law; introduce a 1-round algorithm for approximate point placement in the plane in an adversarial model; and study the relation of neighborhood graphs as subsets of Delaunay triangulations to locally minimum triangulations.

We thank all of the reviewers for their diligence in making recommendations and evaluating revised versions of the papers presented in this TCS journal issue. We would also like to thank all of the authors for submitting their papers to the journal and the associate editors for their valuable work.

It is our hope that this collection of seven articles presented in this special issue will be a valuable resource for Transactions on Computational Science readers and will stimulate further research in the vibrant area of computational science theory and applications.

June 2018   Marina L. Gavrilova
                                                              C. J. Kenneth Tan

LNCS Transactions on Computational Science – Editorial Board

Contents

A Delta-Diagram Based Synthesis
for Cross Layer Optimization Modeling
of IoT

Prathap Siddavaatam and Reza Sedaghat[(✉)]

OPRA-Labs, Ryerson University, 350 Victoria Street,
Toronto, ON M5B 2K3, Canada
{prathap.siddavaatam,rsedagha}@ee.ryerson.ca
https://www.ee.ryerson.ca/opr/

Abstract. Internet of Things is a networking platform where billions of
every day devices communicate intelligently making every day communication highly informative. The IoT defines a world-wide cyber-physical
system with a plethora of applications in the fields of demotics, e-health,
goods monitoring and logistics, among others. The use of cross-layer
communication schemes to provide adaptive solutions for the IoT is
motivated by the high heterogeneity in the hardware capabilities and
the communication requirements among things. In this article, a novel
Delta Diagram synthesis for the IoT is proposed to accurately capture
both the high heterogeneity of the IoT and the impact of the Internet as part of the network architecture. Furthermore, a novel modified
Grey Wolf Optimizer framework is proposed to obtain optimal routing
paths and the communication parameters among things, by exploiting
the interrelations among different layer functionalities in the IoT. Moreover, a cross-layer communication protocol is utilized to implement and
test this optimization framework in practical scenarios. The results show
that the proposed solution can achieve a global communication optimum
and outperforms existing layered solutions. The novel Delta-diagram is
a preliminary step towards providing efficient and reliable end-to-end
communication in the IoT which may be extended to other dimensions
of IoT like security and hardware synthesis.

Keywords: Internet of Things (IoT) · Network architecture
Synthesis · Wireless Sensor Networks · Delta diagram
Grey Wolf Optimizer · Cross-layer optimization

1 Introduction

Over the past decade Wireless Sensor Networks (WSNs) have been the subject
of intensive research in cross-layer optimization. WSNs comprise a vast number
of sensor nodes deployed for collecting surrounding context and environment

© Springer-Verlag GmbH Germany, part of Springer Nature 2018
M. L. Gavrilova and C. J. K. Tan (Eds.): Trans. on Comput. Sci. XXXIII,
LNCS 10990, pp. 1–24, 2018.
https://doi.org/10.1007/978-3-662-58039-4_1

information [2]. A myriad of medium access and routing protocols [33,52] along with many physical layers have been proposed for WSNs. These kind of protocols have been made energy-aware [58,62]; fusion and aggregation strategies were employed [43]; location enriched basic infrastructures have been deployed [23], timing [66], and security protocols [26] enhanced; operating systems were specifically designed to support such higher-level abstractions [19]. Today, we stand at the cusp of Internet of Things (IoT), which is expected to massively span across these interconnecting WSNs.

The vast heterogeneousness in capacities of Internet of Things (IoT) device hardware and the diverse communication methods used among things warrant a cross-layer based design [4,28] for IoT. The IoT aims to network various things, e.g., smart watches, phones, tablets, sensors, actuators, cars and other mobile devices, utilizing a variety of existing physical network infrastructures, including WiFi, Bluetooth, 802.15.4, Z-wave, and LTE-Advanced [6,15]. Basically, IoT is realized by amalgamating communication capabilities with functionalities of sensing, identification and actuation into everyday things and then initiating communication in expansive Internet technologies.

The joint optimization of the transmission strategies at the various communication stack layers is referred to as cross-layer optimization [28,59]. Recently, there has been tremendous effort toward cross-layer optimizations for wireless communication and WSN [2] protocols. Effectively these optimizations focus on data generated by one layer to optimally make decisions for a different layer in OSI stack. The design based on a cross-layer protocol can reduce a huge number of control messages. Thus improving decision making within the protocol by subsequently pig-tailing control information on ordinary data packets and also organizing and sharing that particular control information with all protocol components.

1.1   Organization

We believe that the proposed research work is the first optimization framework in IoT that is completely adaptable to the dynamic QoS requirements and applicable to any of the layers being considered for cross-layer synthesis. This article is organized as follows: Related work in terms of latest standardization attempts for IoT in Sect. 1.2, an outline of the contributions made by this article in Sect. 1.3, Previous work in terms of the example IoT protocol used for practical scenarios is described in Sect. 1.4. In Sect. 2, we describe our multi-objective formulation utilizing the Delta-diagram synthesizer and develop the cross-layer optimization model which captures the inter-relating parameters among the different layers of the protocol stack. In Sect. 6, we describe the optimizer framework needed to find the near-optimal solution to the multi-objective solution space. In Sect. 7, we evaluate the performance of the proposed solution and the paper is concluded in Sect. 8.

1.2   Related Work

Several cross-layer approaches [16,30,40] for IoT network optimization have been proposed to overcome their limitations. Typically, these proposals aim to minimize the energy consumption, increase routing efficiency, and improve QoS provisioning [47]. The proposals include cross-layer optimization of two layers as well as three layers for the IoT protocol stack [47]. In [60], an algorithm for a cross-layer routing protocol called power control based directed spanning tree (PCDST), which is based on the traditional spanning tree (ST) routing protocol and subsequently reduces the total energy consumption of the network. The network throughput can also be improved using cross-layer optimization of application, media access control (MAC) and physical layers as in [40] to achieve a delay-aware framework. The authors in [42] present a cross-layer framework for IoT communication that employs cognitive radio communication and supports QoS for smart grid applications. They have presented a suboptimal distributed control algorithm (DCA) to support QoS through dynamic spectrum access, flow control, scheduling and routing decisions. Besides being used for optimization purposes, cross-layer approaches [58] have also been used for analysis. In [28] the authors make a cross-layer analysis of three error control schemes: forward error correction (FEC), automatic repeat request (ARQ), and hybrid ARQ. This cross-layer analysis pertains to multi-hop routing, energy consumption, and end-to-end latency. The result describe that FEC and hybrid ARQ schemes are suitable for delay sensitive traffic in WSNs for IoT [28].

We briefly list some of the recent attempts at standardization of point-to-point communications for IoT that are only relevant to our model in regards to OSI protocol [64] stack.

IoT Data Link Layer. The most widely used data link layer standards are Bluetooth Low Energy (BLE) [29,56] and ZigBee [22]. IEEE 802.11ah [51], on the other hand, is the most convenient standard employed due to existing and widely separated infrastructure of IEEE 802.11 [27]. Also, IEEE 802.15.4 is the most commonly used IoT standard for medium access control (MAC) sublayer of data link layer. However, some providers would seek for more reliable and secured technology and have recently employed HomePlug [41] for LAN connectivity. In addition, the newly arising LoRaWAN [1] has made huge strides in usage of low-power Wide Area Networks (WANs) having low cost, mobility, security, and bi-directional data link communication for IoT applications.

IoT Network Layer. Typically the network layer is classified into two sub-layers a routing layer which handles the transfer of point-to-point packets from source to destination, and an encapsulation layer that forms the packets. IPv6 over Low power Wireless Personal Area Network (6LoWPAN) [11] is the most commonly used encapsulation protocol for IPv6 in IoT MAC frame. It efficiently encapsulates IPv6 long headers in IEEE802.15.4 [18] small packets restricted to a maximum of 128 bytes. IPv6 over Networks of Resource-constrained Nodes (6Lo) [31] is another set of standards for transmission of IPv6 frames on various data links. This encapsulation protocol standard was developed recently by a working group in IETF [31].

IoT Physical Layer. The physical layer is primarily responsible for establishing a reliable and physical link to transmit binary data. Low Power Wide Area Network (LPWAN) [48] physical layer technologies include LoRa (Long Range physical layer protocol), Haystack [45], SigFox [14], Long-Term Evolution (LTE)-M, and NB-IoT (Narrow-Band IoT) [55]. The LPWAN NB-IoT and LTE-M standards are aimed at providing low-power, low-cost IoT communication options using existing cellular networks. Ethernet implements the IEEE 802.3 physical layer standard that is widely deployed for wired connectivity within local area networks. A few of the IoT devices are designed to remain stationery. For example, sensor units that are installed within a building automation system can use wired networking technologies like Ethernet. Several physical layers standards have been proposed recently based on application specific requirements as discussed in [57].

1.3 Contribution

In this article, we focus on developing a novel Delta-Diagram based synthesizer model for cross-layer optimization tailored to suit individual optimization strategies. These strategies depend on the diverse synthesizing requirements of the IoT network depending on behavior, structure and optimizer domain based parameters considered at each individual layer. The synthesizing layout of Delta-Diagrams that are relevant for our study in this article are illustrated in Figs. 2a, b and c with respect to Data Link, Network and Physical layers. Furthermore, our approach is capable of withstanding uncertain headwinds arising in IoT due to differing hardware capabilities and communication requirements among different types of things. The proposed cross-layer solution using Delta-Diagram synthesizer is widely adaptable for any combination of the protocol stack layer based parameters. The choice of the layers is specifically driven by the IoT network optimization requirements and parameters chosen by the multi-objective model resolving trade-off between QoS requirements. The contributions of this article can be summarized as follows,

Synthesizer. A novel Delta-Diagram synthesizer that is a tool by which an abstract form of required communication layer behavior, typically at receiver-transmitter level, is turned into a design implementation in terms of packet structure and optimizer characteristics.

Universality. The proposed cross-layered approach to synthesizing a communication system is applicable to a plethora of optimization scenarios in IoT communication depending on the targeted protocol stack layers and device capabilities.

Multi-objectivity. A novel multi-objective function formulated for specific IoT case study adopted to demonstrate the viability of the Delta-Diagram synthesizer and cater to different QoS requirements, ranging from error-limited or energy-depraved applications to hyper-sensitivity delay applications or any type of combination of them subjected to specific constraint of applications.

Optimizer. A novel modified Grey Wolf Optimizer (MGWO) algorithm to solve the multi-objective function discussed in the previous point. The MGWO algorithm makes iterations of investigation of new regions and exploiting solutions until reaching near-optimal solution.

Performance. A new theoretic point-to-point (p2p) cross-layer optimization framework is proposed for IoT that provides a systematic, rather than ad hoc, mechanism for dynamically selecting and adapting the transmission strategy at each layer and the message exchange across layers. The performance was vastly superior compared to contemporary cross-layer approach [44] for IoT and is demonstrated by simulation results.

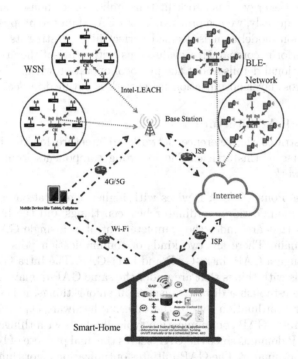

Fig. 1. Network architecture of delta diagram synthesized IoT

1.4   IoT System Overview

1.4.1   Protocol Set-Up

Although several cross-layer protocols have been devised recently for Wireless Sensor Networks (WSNs) [37], Wireless Mesh Networks (WMNs) [17] or Ad Hoc Networks (AHNs) [21], they cannot be utilized for synthesizing and experimenting the proposed framework for IoT. This is due to the fact that they are incapable of capturing the heterogeneity of the IoT which usually incorporates things that have contrasting hardware capabilities, QoS constraints and data precision requirements. Contrarily, the listed protocols in [37], [9,21] cater to WSNs whose nodes usually have very similar hardware specifications, common communication requirements and shared objectives. In addition, the Internet is involved in the IoT network reference architecture mimics practical scenarios that are usually centralized and hierarchically organized unlike the flat networks without Internet proposed in [37], [17,21]. In [62], we have developed the cross-layer protocol called *intel-LEACH* which specifically integrates physical, MAC as well as routing layer functionalities into a unified communication framework. Furthermore, the optimized protocol intel-LEACH selects high yield nodes based on a dynamic optimization strategy independent of the network architecture layout. The protocol intel-LEACH also guarantees higher data precision with maximal residual energy vs network lifetime unlike other non-extensible WSN protocols. Consequently, we use protocol intel-LEACH on the proposed IoT network optimization model as it is deemed perfect fit for testing its performance metrics and perform comparative evaluations with state-of-the-art cross-layer protocols. Additional details about the protocol operation within the context of the proposed cross-layer optimization framework is outlined in Sect. 6.

1.4.2   Network Architecture

The Fig. 1 illustrates the reference network architecture for the IoT utilized during the course of this article. The following components form an integral part of the network,

Gateway Access Point (GAP): Devices with higher computation capacity that acts as a communication coordinator between things and the Internet. The set of things that are under the complete control of a single GAP is known as GAP domain. There are two kinds of communication possibilities while routing through a GAP; intra-GAP and inter-GAP. The Intra-GAP communication deals with things that are under the same GAP domain. Although it is possible to establish a direct transmission among things, it is advisable to use GAP for coordination due to the differing hardware capabilities among things. The Inter-GAP communication typically deals with things that are in different GAP domains since the GAP serves the dual purpose of being a gateway and a coordinator. The GAP initiates optimization algorithms locally by utilizing its network knowledge. e.g. Wi-fi router in Smart Home network, Bluetooth router in BLE-network and cluster heads in WSN comprise the GAPs in Fig. 1.

Things: Devices or objects with differing capabilities due to diverse hardware specifications in terms of communication, computation, memory and data storage capacity, or transmission power. As depicted from Fig. 1, smart homes with communicating home appliances and other sorts of equipment, are some examples of things. All the things do not have the capability to directly communicate with the GAP, in which case multi-hop links are required. Contrarily, the GAP is able to directly communicate with all the things under its domain in a single hop. This asymmetry in the links results arises due to the higher transmission power of the GAP as compared to many of the things.

Internet: It is the rudimentary part of the IoT. While synthesizing the network model using Delta-diagram, the Internet will be considered as fundamental component that is characterized by an stochastic queuing delay model and a stochastic packet loss model [50]. As shown in Fig. 1 a plethora of things (smoke alarms, refrigerators etc.,) are connected to the Internet via a common Gateway Access Point (GAP).

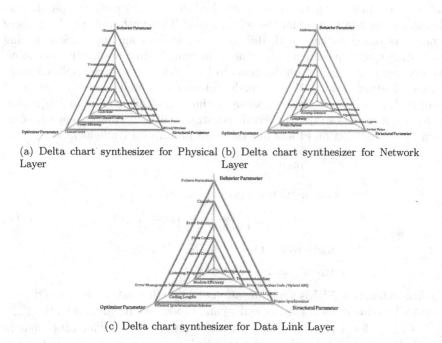

(a) Delta chart synthesizer for Physical Layer

(b) Delta chart synthesizer for Network Layer

(c) Delta chart synthesizer for Data Link Layer

Fig. 2. Delta diagram synthesizers

2 Cross-Layer Design Synthesis

In this section, we propose a novel cross-layer optimization model for the IoT using a parameter synthesizing road map provided by Delta diagrams as illustrated in Figs. 2a, b and c for physical, network and data link layers respectively. Furthermore the Delta diagram synthesizer can be generalized and extended to any layer chosen by the optimization model adopted by a communication system. For the IoT network architecture specified in Fig. 1, the Delta diagram is applied to synthesize three layers: Physical, network and data link layers of the protocol stack described in Fig. 3. We adopt a resource allocation approach which involves centralized management [2] to estimate resource availability and environmental dynamics, coordinate the allocation of resources across applications and nodes, and therefore adapt the protocol parameters at each layer based on the synthesizer road map provided by Delta diagram. This approach assists in integrating scattered communication functionalities into a united coherent optimization model and provide an flexible solution for cross-layer design and control. We use this approach to jointly control and synthesize a select case of Quality of Service (QoS) requirements. Based on our selected QoS case, we synthesize the physical layer of network architecture of Fig. 1 using the Delta diagram in Fig. 4 that results in parameters, channel and modulation. Similarly, data link layer synthesizer (Fig. 5a) utilizes only MAC layer parameters like Access Control and error control and network layer synthesizer in Fig. 6 produces parameters like addressing and routing for the selected case. The synthesizer produces these parameters based on case with differentiated services for applications having contrasting QoS requirements, ranging from error-limited applications or minimum energy consumption applications to highly-delay-sensitive applications or any combination of them. We can model this case as a multi-objective optimization problem that must simultaneously optimize multiple conflicting objectives of QoS requirements subject to certain constraints of Delta diagram synthesizer, given by a minimization fitness function that produces a triple as follows,

$$\mathcal{F}^{\text{fit}} = \underset{x}{\text{minimize}}\; J(\mathsf{x})$$

$$\iff \underset{x}{\text{minimize}}\; J(\mathsf{x}) = \left(\; x_0,\; x_1,\; x_2\; \right)$$

$$= \left(\xi_{min}^{\text{p2p}}, \mathcal{T}_{min}^{\text{p2p}}, \text{PE}_{min}^{\text{p2p}}\right) \tag{1}$$

$$\text{subject to } \delta^\xi \cdot O_1 \cdot \delta^{\text{PE}} + O_2 \cdot \delta^{\text{PE}} + O_3 \cdot \delta^{\mathcal{T}}$$

$$\text{where } \delta^{\text{PE}} = 1 - (\delta^\xi + \delta^{\mathcal{T}})$$

The fitness function $\mathcal{F}^{\text{fit}}$ is a minimization function. It takes a vector x of dimensions 3×1 having elements x_0, x_1 and x_2 and produces a minimized triple (ξ_{min}^{p2p}, $\mathcal{T}_{min}^{\text{p2p}}$, $\text{PE}_{min}^{\text{p2p}}$) for a given instance when subjected to constraint conditions in Eq. (1). For the IoT network architecture in Fig. 1, we consider the end-to-end communication as point-to-point communication due the vast diversity of end

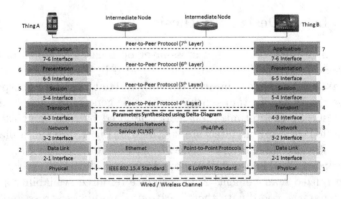

Fig. 3. Open Systems Interconnection reference model magnifying Network, Data Link and Physical layers

nodes and its heterogeneousness with respect to its computing capacity, storage, etc. The terms in Eq. (1) are as follows: PE^{p2p} is point-to-point packet error rate ratio which is the number of error packets after applying Forward Error Correction (FEC) divided by the total number of received packets. It is important to note that a packet is the unit of data for radio transmission with applicable FEC. $\mathcal{T}^{p2p}$ and ξ^{p2p} represent point-to-point delay and energy consumption, while δ^{ξ}, δ^{PE}, $\delta^{\mathcal{T}}$ are linear weighting coefficients for energy consumption, time delay and point-to-point packet error rate respectively. PE^{opt}, $\mathcal{T}^{Opt}$ and ξ^{opt} are defined as the quintessential values [46] for point-to-point energy consumption, delay and packet error rate for normalizing purposes respectively. These quintessential values are defined to be the most optimistic and usually unattainable minimum values, which are used to provide the non-dimensional objective functions and can be computed off line. These values are used to normalize each of constraint terms O_1, O_2 and O_3, and optimize their deviations with respect to a pre-defined threshold, instead to minimize their absolute values. This is done due to the fact that there are three differing objectives which are measured in separate units as well as their order of magnitude. The constraints terms in Eq. (1) are defined as follows,

- Point-to-point energy coefficient is

$$O_1 \leftarrow |\frac{\xi^{opt} - \xi^{p2p}}{\xi^{opt}}| \cdot |\frac{PE^{p2p} - PE^{Th}}{PE^{Th}}| \qquad (2)$$

- Point-to-point packet error rate coefficient is

$$O_2 \leftarrow |\frac{PE^{p2p} - PE^{opt}}{PE^{opt}}| \qquad (3)$$

- Point-to-point timing coefficient is

$$O_3 \leftarrow |\frac{\mathcal{T}^{p2p} - \mathcal{T}^{Opt}}{\mathcal{T}^{Opt}}| \qquad (4)$$

3   Physical Layer Synthesis

At the level of physical layer (Fig. 3), the things of IoT can have their communication synthesized on any of the network domains relating to Behavioral, structural and Optimizer parameters depicted in Fig. 2a. For example, in case of network architecture of Fig. 1, different things have a variety of maximum transmission power parameter which tend to choose differing modulation schemes and have varying data storage capacities (i.e., the number of packets that things can locally queue). As a consequence the physical layer Delta diagram synthesizer of Fig. 4 depicts this action corresponding to parameter level parsing across specified domains.

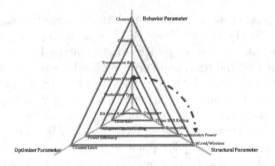

Fig. 4. Parameter synthesizing action for physical layer synthesis

3.1   Channel Medium and Frequency Spectrum

Our model adopts the frequency spectrum allocation based on IEEE 802.15.4 standard [3] where things are able to dynamically select any of the 5-MHz-wide sub-channels in 2400–2480 MHz band. Due to the fact that many applications of IoT are based indoor (e.g., home, office, warehouses), we set the ITU channel model for indoor propagation [8]. The total path loss $\mathcal{L}$ in dB is given by

$$\mathcal{L}(d, f_c) = A \log_{10}(d) + \mathcal{L}_f(r) + 20 \log_{10}(f_c) - 28 \tag{5}$$

where f_c represents carrier frequency in MHz, d is the distance in meters to transmit, r is the number of floors between the transmitter and the receiver (we consider only one floor in our current scenario, i.e., $r = 1$ and A is the distance attenuation coefficient (i.e., $A = 20$ in our simulations), and $\mathcal{L}_f$ is the floor penetration loss factor since whose value is zero for only one floor.

3.2   Power, Modulation and BE for Transmission

The Bit Error Rate (BE) is directly affected by the transmission power and modulation. The Bit Error Rate over a data link ℓ denoted by $\mathsf{BE}_\ell^{\mathrm{DL}}$ is determined by its respective Signal-to-Noise Ratio $\mathsf{SNR}_\ell^{\mathrm{DL}}$ and modulation M_ℓ as,

$$\mathsf{BE}_\ell^{\mathrm{DL}}(d, \ f_c) = \mathcal{S}_\psi \Big(\mathsf{M}_\ell, \ \mathsf{SNR}_\ell^{\mathrm{DL}}(d, \ f_c) \Big)$$
$$\text{where } \mathsf{SNR}_\ell^{\mathrm{DL}}(d, \ f_c) = \mathsf{P}_\ell^{Tx} - \mathsf{P}^{\mathrm{noise}} - \mathcal{L}_\ell^{\mathrm{DL}}(d, \ f_c) \tag{6}$$

where the BE is calculated by function $\mathcal{S}_\psi$ for a given modulation and SNR, and it is common knowledge for standard modulations. From Eq. (6), we observe that the SNR (in dB) of a data link ℓ has a transmission power P_ℓ^{Tx} in dB, total noise power $\mathsf{P}^{\mathrm{noise}}$ in dB at the receiver and a path loss $\mathcal{L}_\ell^{\mathrm{DL}}(d, f_c)$ for data link ℓ estimated using Eq. (5)

The predominant modulations namely, BPSK, QPSK and 16-QAM are deployed for simulation purposes using intel-LEACH protocol [62], but any other modulation can be used in our framework. The achievable BE for any given SNR are distinguishable for these specified modulations i.e., $\mathcal{S}_\psi$ in Eq. (6), and spectral efficiency, i.e., Given a transmission bandwidth, it specifies the theoretically achievable maximum data bit-rate.

3.3   Probability of Packet Drop-Out and Data Buffer Capacity

The probability of packet drop-out is directly influenced by data storage capacity μ of the things. Thus, the probability of discarding a packet at link ℓ, $\mathbb{P}_\ell^{\text{drop}}$ is related to the fact that it cannot be queued at the transmitter or at the receiver given by,

$$\mathbb{P}_\ell^{\text{drop}} = \mathcal{G}_\psi(\mu_\ell, \Re_\ell) \tag{7}$$

such that the maximum number of packets μ_ℓ that can be queued at the transmitter or the receiver and the total local traffic $\Re_\ell$ (self traffic and relayed traffic) is influenced by function $\mathcal{G}_\psi$. Considering a simplest case of Poisson traffic, the transmitter and receiver are modeled as single serve First In First Out (FIFO) queue having data buffer capacity of μ_ℓ.

4   Data Link Layer Synthesis

In IEEE 802 LAN/MAN standards, the medium access control (MAC) sublayer (also known as the media access control sublayer) and the logical link control (LLC) sublayer together make up the data link layer. Within that data link layer, the LLC provides flow control and multiplexing for the logical link (i.e. EtherType, 802.1Q VLAN tag, etc.), while the MAC provides flow control and multiplexing for the transmission medium.

For the IoT network architecture in Fig. 1, the delta diagram for data link layer synthesizer (Fig. 2b) utilizes only MAC layer parameters like Access Control and error control as per QoS requirements defined for Eq. (1). We analyze the impact of the error control mechanism and the MAC protocol on the network performance, as well as, their interrelations with other layers and the limitations imposed by the things capabilities. This parameter level parsing action is depicted sequentially in data link layer delta diagram synthesizer of Figs. 5a, b and c.

4.1   Packet Error Rate and Error Control

For our cross layer model we adopt a Hybrid ARQ error control scheme [67], the overall packet error rate over data link ℓ is given by

$$\text{PE}_\ell^{\text{overall}} = \mathcal{K}_\psi\left(\text{PE}_\ell^{\text{non-Code}}, \text{N}_\ell^{\text{max}}, \text{N}_{\text{FEC}}^{\text{bits}}\right) \tag{8}$$

where $\mathcal{K}_\psi$ is a function that relates the PE of data link ℓ after Hybrid ARQ error control, $\text{PE}_\ell^{\text{overall}}$, with the uncoded data link $\text{PE}_\ell^{\text{non-Code}}$. In the above equation, $\text{N}_{\text{FEC}}^{\text{bits}}$ is the FEC redundancy length and $\text{N}_\ell^{\text{max}}$ is the maximum number

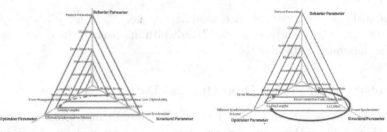

(a) Parameter synthesizing start action for Data Link Layer Synthesis

(b) Parameter synthesizing sophomore action for Data Link Layer Synthesis

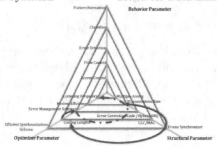

(c) Parameter synthesizing final action for Data Link Layer Synthesis

Fig. 5. Delta diagram data link synthesizers

of transmissions including retransmissions, which can be adjustable in our cross-layer model.

The uncoded packet error rate $\mathsf{PE}_\ell^{\text{non-Code}}$ over a link ℓ in Eq. (8) is given by,

$$\mathsf{PE}_\ell^{\text{non-Code}} = \left[1 - (1 - \mathsf{BE}_\ell^{\text{DL}})^{\mathsf{N}^{\text{bits}}}\right] \\ \cdot (1 - \mathbb{P}_\ell^{\text{drop}}) \tag{9}$$

where N^{bits} is obtained by network layer synthesis calculation specified in Sect. 5

Using Eq. (8), $\mathsf{PE}^{\text{m-Hops}}$ is defined as a function of PE over link ℓ given by,

$$\mathsf{PE}^{\text{m-Hops}} = 1 - \prod_{\ell=1}^{\text{m Hops}} (1 - \mathsf{PE}_\ell^{\text{overall}}) \tag{10}$$

From Eq. (8) we can infer that higher magnitude values of $\mathsf{N}_{\text{FEC}}^{\text{bits}}$ and $\mathsf{N}_\ell^{\text{max}}$ lead to (i) a lower transmission data rate or larger link delay, (ii) a lower Packet Error Rate (PER), and (iii) a higher total energy consumption. The impact of the error control parameters interplays with those made by adjusting the transmission power and the modulation scheme at the physical layer. Finally, the point-to-point packet error rate constricted by the threshold PE^{Th} dependent on $\mathsf{PE}^{\text{m-Hops}}$, and PE^{Net} (PE over the Internet) is defined by,

$$\mathsf{PE}^{\text{p2p}} = 1 - (1 - \mathsf{PE}^{\text{m-Hops}})(1 - \mathsf{PE}^{\text{Net}}) \leq \mathsf{PE}^{\text{Th}} \tag{11}$$

4.2   Medium Access Control

By utilizing a hybrid protocol [61] of Sleep MAC (SMAC) [68] and CSMA/CA medium access protocol, the medium access method can eliminate the interference drastically if the carrier sensing is correctly performed. This results in substantial savings in energy as well as reduce the interference among the things. For example, almost, 90% of the total energy consumed during idle listening can be saved when things use an awake/sleep duty cycle of 10%.

5   Network Layer Synthesis

We discuss the impact of the packet size on the different layers. During the stage of network layer, the things of IoT can have variations based on the Behavioral, structural and Optimizer parameters depicted in synthesizer Fig. 2b. We analyze the addressing and information routing on the network performance, as well as, their interrelations with other layers and the limitations imposed by the things capabilities. The initiation of parameter level parsing action is depicted in network layer delta diagram synthesizer of Fig. 6.

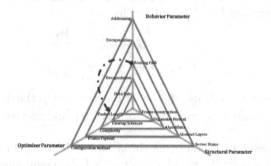

Fig. 6. Parameter synthesizing begin action on Network Layer

5.1   Addressing of Things and Routing

Since a huge amount of things are part of the IoT architecture depicted in Fig. 1, it is imperative to use IPv6 addressing for the IoT. IPv6 addresses are expressed by 128 bits, which allow the definition of 10^{38} unique addresses (these are expectedly enough for the time being). However, IPv6 addresses are only used for inter-GAP communications, while much shorter local addresses are used in intra-GAP communications.

Our model for routing mechanism uses destination-based routing mechanism [7] where the GAP selects the point-to-point route and the configuration parameters for each data link in a centralized manner resulting from cross layer modeling.

5.2   Packet Size Impact

In our model, a fixed packet size $\mathsf{N}^{\mathrm{bits}}$ is selected and used for all the data links ℓ throughout a given path. A larger packet size results in a reduced point-to-point delay by saving the handshake time $\mathcal{T}^{\mathrm{hshake}}$, the acknowledgement time $\mathcal{T}^{\mathrm{ack}}$, and the queuing time $\mathcal{T}^{\mathrm{Queue}}$ that are required for each packet. Additionally, the reduction of the total number of packets to be sent has an impact on the total energy consumption, while at the same time, transmitting more bits of information in a packet affects the Packet Error Rate (PER). All these interrelations are incorporated in our cross-layer framework.

$\mathsf{N}^{\mathrm{bits}} = \mathsf{N}^{\mathrm{bits}}_{\mathrm{Header}} + \mathsf{N}^{\mathrm{bits}}_{\mathrm{FEC}} + \mathsf{N}^{\mathrm{bits}}_{\mathrm{Data}}$ where $\mathsf{N}^{\mathrm{bits}}$ is packet size containing header size, data length, and FEC redundancy length, $\mathsf{N}^{\mathrm{DL}}_{\ell} = (1 - \mathsf{PE}^{\mathrm{non\text{-}Code}}_{\ell})^{-1}$ is the upper bound for the total number of transmissions of a packet with correct decoding over link ℓ. We can now describe the overall energy consumption and timing delays (for complete point-to-point path of the IoT network) having constraints $\eth$ and $\mathcal{T}^{\mathrm{Th}}$ in evaluating the Fitness function in Eq. (1) as follows.

$$
{}^{k}\xi = \mathsf{N}^{\mathrm{bits}} \cdot \xi^{k}_{b}
$$
$$
\Longleftrightarrow \xi^{k} \leq \xi^{k}_{\mathrm{Th}}
\tag{12}
$$

where energy on the k^{Th} node is ${}^{k}\xi$ resulting from the product of packet size $\mathsf{N}^{\mathrm{bits}}$ and the energy per bit ξ^{k}_{b} calculated by,

$$
\xi^{k}_{b} = 2\xi^{\mathrm{Eqp}}_{b} + \frac{\mathsf{P}_{\mathrm{Tx}}}{\Re_{k}}
$$
$$
\Longleftrightarrow \xi^{\mathrm{Eqp}}_{b} = \xi^{\mathrm{Eqp}}_{b-Tx} = \xi^{\mathrm{Eqp}}_{b-Rx} \ \mathrm{J/bit}
\tag{13}
$$

where ξ^{Eqp}_{b} is the energy to transmit one bit independent of the distance involved and $\xi^{\mathrm{Eqp}}_{b-Tx}$, $\xi^{\mathrm{Eqp}}_{b-Rx}$ constitute energy for transmitter and receiver device equipment.

Let $\mathcal{T}^{\mathrm{p2p}}$ be the point-to-point time duration which also includes the Internet delay $\mathcal{T}^{\mathrm{Net}}$ constricted by threshold $\mathcal{T}^{\mathrm{Th}}$ for inter-GAP communications and the link queuing delay $\mathcal{T}^{\mathrm{Queue}}$. Factors such as current traffic, behavior of other nodes in IoT or its hardware status etc., determine the queuing delay and Internet delay. Thus $\mathcal{T}^{\mathrm{p2p}}$ is defined as,

$$
\mathcal{T}^{\mathrm{p2p}} = \sum_{\ell=1}^{\mathrm{m\ Hops}} (\mathcal{T}^{\mathrm{Queue}}_{\ell} + \mathcal{T}_{\ell} + \mathcal{T}^{\mathrm{Net}}_{\ell})
$$
$$
\Longleftrightarrow \mathcal{T}_{\ell} \leq (\mathsf{N}^{\mathrm{DL}}_{\ell} - 1) \cdot (\mathcal{T}^{\mathrm{hshake}}_{\ell} + \mathcal{T}^{\mathrm{Data}}_{\ell} + \mathcal{T}^{\mathrm{Time\text{-}out}}_{\ell})
$$
$$
+ \mathcal{T}^{\mathrm{Sleep}} + \mathcal{T}^{\mathrm{Signal}}(\mathcal{T}^{\mathrm{hshake}} + \mathcal{T}^{\mathrm{Data}} + \mathcal{T}^{\mathrm{ack}})
\tag{14}
$$

Using Central Limit Theorem [13], the complete point-to-point delay $\mathcal{T}^{\mathrm{p2p}}$ can be modeled as a Gaussian distributed random variable with variance given below,

$$
\mathrm{variance} = \mathrm{var}(\mathcal{T}^{Net+Queue}) \text{ where}
$$
$$
\mathcal{T}^{Net+Queue} = \sum_{\ell=1}^{\mathrm{m\ Hops}} \mathcal{T}^{\mathrm{Queue}}_{\ell} + \mathcal{T}^{\mathrm{Net}}
\tag{15}
$$

The target probability t_γ is obtained by Chebyshev's inequality [34] to decompose point-to-point delay constraint into,

$$\mathbb{P}(T^{\mathrm{p2p}} \geq T^{\mathrm{Th}})$$

$$\leq \frac{\mathrm{var}(T^{Net+Queue})}{\mathrm{var}(T^{Net+Queue}) + (T^{\mathrm{Th}} - \sum_{\ell=1}^{m\ \mathrm{Hops}} T_\ell - \overline{T^{Net+Queue}})^2}$$

$$\leq 1 - t_\gamma \qquad (16)$$

$$\iff \mathbb{P}(T^{\mathrm{p2p}} \leq T^{\mathrm{Th}}) \geq t_\gamma \text{ satisfying}$$

$$(T^{\mathrm{Th}} - \sum_{\ell=1}^{m\ \mathrm{Hops}} T_\ell - \overline{T^{Net+Queue}}) > 0$$

The complete point-to-point throughput of IoT architecture is inversely proportional to the point-to-point delay ^{p2p}T given by,

$$\rho^{\mathrm{p2p}} = \frac{\mathrm{N}_{\mathrm{Data}}^{\mathrm{bits}}}{T^{\mathrm{p2p}}} \geq \rho^{\mathrm{Th}} \qquad (17)$$

6   GWO Cross-Layer Optimizer Framework

6.1   Background on Grey Wolf Optimizer

We adopt a GWO framework proposed in [49] and apply it to extract a minimized triple using fitness function defined in Eq. (1) by synthesizing the parameters discussed in delta diagram Figs. 4, 5c and 6. In [49], the core function lies in closely mimicking the behavior of the wolves in prey hunting. Grey wolves follow four steps in searching and hunting for the prey. These include, hunting the prey, encircling it, attacking and eating it and then hunting again. Figure 7b shows a visualization of the search space positioning for the GWO algorithm and follows a strict social hierarchy shown in Fig. 7a.

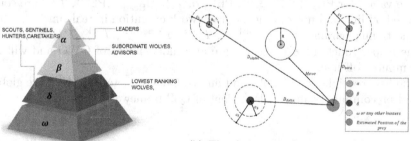

(a) Social hierarchy of Grey Wolves.

(b) The design search space positioning for GWO-DSE [49]

Fig. 7. Grey Wolf Optimizer

6.2   Algorithm Set-Up

For mathematically modeling the algorithm using GWO, we need the basic notations defined:

$$
\begin{aligned}
\overrightarrow{D_\alpha} &= \mid \overrightarrow{C}_1 \times \overrightarrow{X}_\alpha(t) - \overrightarrow{X}(t) \mid \\
\overrightarrow{X_1}(t) &= \overrightarrow{X}_\alpha(t) - \overrightarrow{A}_2(t) \times \overrightarrow{D_\alpha} \\
\overrightarrow{D_\beta} &= \mid \overrightarrow{C}_2 \times \overrightarrow{X}_\beta(t) - \overrightarrow{X}(t) \mid \\
\overrightarrow{X_2}(t) &= \overrightarrow{X}_\beta(t) - \overrightarrow{A_2} \times \overrightarrow{D_\beta} \\
\overrightarrow{D_\delta} &= \left| \overrightarrow{C_3} \times \overrightarrow{X_\delta}(t) - \overrightarrow{X}(t) \right| \\
\overrightarrow{X_3}(t) &= \overrightarrow{X}_\delta(t) - \overrightarrow{A_3} \times \overrightarrow{D_\delta} \\
\overrightarrow{X}(t) &= \frac{\overrightarrow{X}_1 + \overrightarrow{X}_2 + \overrightarrow{X}_3}{3}
\end{aligned}
\tag{18}
$$

where t is the current iteration A and C are coefficients of the vectors and X_α, X_β, X_δ are the locations of the alpha, beta and the delta wolves and remaining X are omega wolves. The coefficients A and C are calculated as:

$$
\begin{aligned}
\overrightarrow{A}(t) &= 2\overrightarrow{a}(t) \times \overrightarrow{r}_1 - \overrightarrow{a} \\
\overrightarrow{C} &= 2\overrightarrow{r}_2
\end{aligned}
\tag{19}
$$

where a's value decreases linearly from 2 to 0 as the iterations proceed and r_1 and r_2 are random numbers ranging in [0,1]. The GWO-Cross-layer framework records fitness values $\mathcal{F}^{\text{fit}}$ using Eq. (1) for each candidate solution $\overrightarrow{X_i}$. The following notations and assumptions are to be conveyed before designing the algorithm for GWO-Cross-layer framework.

- Each and every candidate solution $\overrightarrow{X_i}$ is a triple defined in Eq. (1).
- The three most optimal candidate solutions are shortlisted as $\overrightarrow{X_\alpha}$, $\overrightarrow{X_\beta}$ and $\overrightarrow{X_\delta}$ respectively.
- Every fitness function call is subjected to the QoS constraints of Eqs. (2)–(4).
- The values for PE^{opt}, ξ^{opt}, T^{Opt}, PE^{Th}, $\eth$, T^{Th}, $\text{N}^{\text{bits}}_{\text{Header}}$, PE^{Net}, T^{Net} are computed off line and not during the framework execution in real time.
- The timing parameters t_γ, T^{hshake}, T^{Data}, $T^{\text{Time-out}}$, T^{ack}, T^{Queue}, μ_ℓ and the helper functions $\mathcal{K}_\psi$, $\mathcal{G}_\psi$, $\mathcal{S}_\psi$ are also independent of GWO and will be computed off line.
- The linear weights δ^ξ, δ^T, δ^{PE} and link traffic parameters $\Re_\ell$, N^{DL} are global and precomputed values independent of QoS framework.

Algorithm 1. GWO-Cross-layer Framework

Data: W_n–magnitude of the wolf pack, Iter_{MAX} – total number of iterations more than 0, set of ordered triples $\texttt{Striples} \leftarrow \{ (\ \xi^{\text{p2p}}, \mathcal{T}^{\text{p2p}}, \text{PE}^{\text{p2p}}\ ) \}$, user defined constraints specified in Sec (6.2)

Result: Most Minimum Triple:$(\ \xi^{\text{p2p}}, \mathcal{T}^{\text{p2p}}, \text{PE}^{\text{p2p}}\ ) \leftarrow \mathcal{F}^{\text{fit}}(\overrightarrow{X_i})$, Optimal Grey wolf position within given Cross-layer Architecture.

1 **begin**
2   Randomly initialize W_n number of candidate $\overrightarrow{X_i}$ Grey wolves
3   Using Eq. (1) identify the best three triples as X_α, X_β and X_δ wolves
4   Set $t := 0$
5   **repeat**
6   **foreach** $\overrightarrow{X_i} \in \texttt{Striples}$ **do**
7   Update $\overrightarrow{X_i}$ using Eq. (18)
8   **end**
9   Compute Eqs. (19) to update a, A, C
10   **forall the** *Wolves* $\overrightarrow{X_i}$ *of* $\texttt{Striples}$ **do**
11   newPositions $\leftarrow$ get positions using Eq. (18)
12   **if** *newPositions* $\notin$ *Range* **then**
13   newPositions $\leftarrow$ get positions using last equation in Eq. (18)
14   **end**
15   **end**
16   **foreach** $\overrightarrow{X_i} \in \texttt{Striples}$ **do**
17   Deduce $\texttt{Minimum-Triple}$ using fitness function $\mathcal{F}^{\text{fit}}(\overrightarrow{X_i})$ subject to QoS constraints of Eqs. (2), (3) and (4)
18   **end**
19   Rank α, β, and δ positions as premier 3 best solutions based on $\texttt{Minimum Triples}$ from Previous Step.
20         $t \leftarrow t + 1$
21   **until** $t \leq \text{Iter}_{MAX}$
22   Choose $\texttt{Optimal Grey Wolf}$ position given by Eq. (1)
23 **end**

The outcome of Algorithm 1 will be near optimal set of triples which satisfy the minimization requirements of Eq. 1 subjected to QoS requirements of Eqs. (2)–(4).

7 Results and Analysis

The intel-LEACH protocol [62] was incorporated to test our framework from synthesis to optimization steps in practical point-to-point IoT scenarios. The operation is briefly listed in following phases.

1. Transmission Phase
   - Check route validity from point to point nodes and following initialization routines of MAC operation described in Sect. 4.
   - Failure mitigation by generating Route Request (RR) packet containing the destination thing ID directed towards nearest GAP.
2. Service Phase
   - GAP transmits its ID in broadcast mode periodically
   - GAP has sufficiently large power to directly communicate with every thing in its domain (Sect. 1.4.2).
   - The devices register themselves to GAP with Network Attached Storage (NAS) packet.

3. Messaging Phase
   – Upon receiving packet data, the thing sends a Route Acknowledge (RA) packet to the previous hop in the route to show its alive.
   – The above process is repeated in multi-hops scenario until source is reached.
   – Data is transmitted by following the optimal route with the chosen communication parameters defined by delta diagram in Fig. 5c and according to description in Sect. 5.
   – Computation complexity is shifted from things to GAP which reduces multi-flow problems.
4. Routing Phase
   – GAP receives the RR packet via several paths, the intermediate nodes are earmarked as priority candidates for data transmission
   – GAP initiates the GWO-Cross-layer framework for potential paths and QoS requirements to find the near optimal path and the associated communication parameters, as explained in Sect. 6.

The performance of the GWO-Cross-layer framework and traditional layered solutions was compared and analyzed by setting the network with following random variables for testing cross-layer functionalities.

– Additive white Gaussian noise (AWGN) at each link ℓ as $N_\ell \sim \mathcal{N}(\mu_{\text{noise}}, \mathbb{P}^{\text{noise}})$ where $\mu_{\text{noise}} = 0$, $10 \log_{10} \mathbb{P}^{\text{noise}} = -86 \, \text{dB}$
– Queuing Delay for things at each link ℓ as $\mathcal{T}_\ell^{\text{Queue}} \sim \mathcal{N}(10, 10^4) \, \text{ms}$
– Delay due to Internet as $\mathcal{T}^{\text{Net}} \sim \mathcal{N}(10^2, 10^4) \, \text{ms}$
– Packet drop-out rate at every link ℓ as $\mathbb{P}_\ell^{\text{drop}} \sim \mathcal{U}(0, 10^{-1})$
– Packet Error rate of internet as $\text{PE}^{\text{Net}} \sim \mathcal{U}(0, 10^{-4})$

The traditional layered approach has each layer autonomously optimized based on Dynamic programming approach as described in [20,24]. We consider the results only when the QoS is focused on either point-to-point delay minimization, energy consumption minimization, or a linear combination of both, while the threshold for PE is constricted to be less than $\text{PE}^{\text{Th}} = 10^4$. However, for simulation purposes, the total amount of data to transmit per transmission is set to 105 bits and the possible packet sizes N^{bits} are 200, 500 or 2000 bits. It is important to note that for every transmission, the thing randomly selects its destination and there is an equal probability that links are either inter-GAP or intra-GAP. The things are randomly assigned diverse hardware capabilities, in terms of computing, memory, energy storage, power and communication. Consequently the link rate of things ranges among $[250, 10^3, 10^4]$ kbps and the power of things varies among $[10, 30, 50, 80, 100] \, \text{mW}$.

Notice that the error bars in Figs. 8a, b and c represent the uncertainty interval at the 95% confidence level. The four layered solutions that are plotted for comparison differ in the modulation scheme and the packet size, and make use of the shortest path routing. The Fig. 8a illustrates point-to-point delay in ms and the energy consumption in mJs (micros Joules) plotted against the number of things in an IoT network. The distance between the transmitter and the

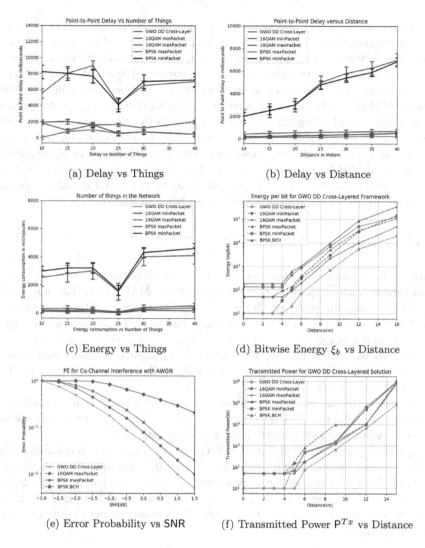

(a) Delay vs Things

(b) Delay vs Distance

(c) Energy vs Things

(d) Bitwise Energy ξ_b vs Distance

(e) Error Probability vs SNR

(f) Transmitted Power P^{Tx} vs Distance

Fig. 8. GWO Delta diagram synthesized cross-layer solution compared with traditional layered solutions

receiver is set to 40 m. Our GWO Delta diagram based cross-layer approach has at least 10% gain over other layered solutions. Consequently, we can observe that neither the point-to-point delay nor the energy consumption increases as the number of things increases. This happens when higher node density creates more optional paths for transmission, and contrastingly the point-to-point consumption does not have proportional relationship with it. The 95% confidence interval implies that our GWO Delta digram Cross-layer solution stabilizes than other layered solutions, and the point-to-point performance does not significantly deviate although the computation complexity at the GAP increases.

Another scenario was observed that by increasing the number of hops in the path implied the rise of the point-to-point delay since there are additional handshake, processing and queuing delay introduced into the transmission and the trend is seen in T^{p2p} increases as distance increases (Fig. 8b). In addition, the energy consumed for the longer distance and by the additional nodes increase the overall ξ^{p2p} (Fig. 8c). It can be inferred from Fig. 8d that the energy consumed per bit (Eqs. 12 and 13) is directly proportional to the distance between point to point and also evident that energy consumed by GWO Delta diagram cross-layer model is lower than referenced methods. Also PE improves significantly with increased SNR for our model compared to other benchmarks as depicted in Fig. 8e. The transmission power is very much lower compared to other layered approaches involving the distance covered as shown in Fig. 8f.

8 Conclusion

In this work, we presented the novel cross-layer design of the IoT protocol stack, which ranges bottom-up from the Physical to the MAC layer. The approach uses a Delta-diagram based synthesizer to identify the parameters to consider for optimization. The parameters chosen are completely dependent on the type of network and topology of IoT setup and hence it is very flexible. Furthermore, synthesis process enables parsing differing levels of parameters and moving across different domains of Behavior, Structural and Optimizer requirements of the chosen protocol for IoT.

We constructed a model for deriving the fitness function which simultaneously minimizes Energy, Timing delays and Packet Error rate requirements of the IoT network protocol subjected to the QoS constraints.

We introduced a modified Grey Wolf Optimizer algorithm to optimize and search the near optimal minimized triple resulting from model search space. The results and analysis section shows that our approach using a test protocol outperforms other cross-layered approaches significantly and is highly flexible in nature so that it can adapted to any type QoS requirements of different IoT network protocols.

References

1. Sinha, R.S., Wei, Y., Hwang, S.H.: A survey on LPWA technology: LoRa and Nb-IoT. ICT Express **3**(1), 14–21 (2017)
2. Sah, D.K., Amgoth, T.: Parametric survey on cross-layer designs for wireless sensor networks. Comput. Sci. Rev. **27**, 112–134 (2018)
3. IEEE 802.15: IEEE 802.15 wireless personal area networks task group 4, January 2016. http://www.ieee802.org/15/pub/TG4.htm
4. Akyildiz, I.F., Vuran, M.C.: XLP: a cross-layer protocol for efficient communication in wireless sensor networks. IEEE Trans. Mob. Comput. **9**, 1578–1591 (2010)
5. Akyildiz, I.F., Wang, X.: Cross-layer design in wireless mesh networks. IEEE Trans. Veh. Technol. **57**(2), 1061–1076 (2008)

6. Al-Fuqaha, A., Guizani, M., Mohammadi, M., Aledhari, M., Ayyash, M.: Internet of Things: a survey on enabling technologies, protocols, and applications. IEEE Commun. Surv. Tutor. **17**(4), 2347–2376 (2016)
7. Al-Karaki, J.N., Kamal, A.E.: Routing techniques in wireless sensor networks: a survey. IEEE Wirel. Commun. **11**(6), 6–28 (2004)
8. Almesaeed, R., Ameen, A.S., Doufexi, A., Dahnoun, N., Nix, A.R.: A comparison study of 2D and 3D ITU channel model. In: 2013 IFIP Wireless Days (WD), pp. 1–7, November 2013
9. Ammar, A.B., Dziri, A., Terre, M., Youssef, H.: Multi-hop leach based cross-layer design for large scale wireless sensor networks. In: 2016 International Wireless Communications and Mobile Computing Conference (IWCMC), pp. 763–768, September 2016
10. Aslani, Z., Aijaz, A.: COOP-RPL: a cooperative approach to RPL-based routing in smart grid AMI networks. CoRR abs/1706.05134 (2017)
11. Bormann, C., Toutain, L., Cragie, R.: IPv6 over low-power wireless personal area network (6LowPAN) routing header (2017)
12. Buonaccorsi, N., Cicconetti, C., Mambrini, R., Podias, N., Russell, P.: ETSI M2M release 1 demonstration. In: 2012 IEEE International Symposium on a World of Wireless, Mobile and Multimedia Networks (WoWMoM), pp. 1–3 (2012)
13. Cam, L.L.: The central limit theorem around 1935. Stat. Sci. **1**(1), 78–91 (1986)
14. Centenaro, M., Vangelista, L., Zanella, A., Zorzi, M.: Long-range communications in unlicensed bands: the rising stars in the IoT and smart city scenarios. IEEE Wirel. Commun. **23**(5), 60–67 (2016)
15. Chai, F., Zhu, T., Kang, K.D.: A link-correlation-aware cross-layer protocol for IoT devices. In: 2016 IEEE International Conference on Communications (ICC), pp. 1–6, May 2016
16. Chze, P.L.R., Leong, K.S., Wee, A.K., Sim, E., May, K.E., Wing, H.S.: Cross-layer secured IoT network and devices. In: Handa, H., Ishibuchi, H., Ong, Y.-S., Tan, K.-C. (eds.) Proceedings of the 18th Asia Pacific Symposium on Intelligent and Evolutionary Systems - Volume 2. PALO, vol. 2, pp. 319–333. Springer, Cham (2015). https://doi.org/10.1007/978-3-319-13356-0_26
17. Conti, M., Maselli, G., Turi, G., Giordano, S.: Cross-layering in mobile ad hoc network design. Computer **37**(2), 48–51 (2004)
18. Culler, D.E., Hui, J.: 6LowPAN tutorial IP on IEEE 802.15.4 low-power wireless networks (2007)
19. Dixon, C., et al.: An operating system for the home. In: Proceedings of the 9th USENIX Conference on Networked Systems Design and Implementation, NSDI 2012, pp. 25–25. USENIX Association, Berkeley (2012)
20. Dong, Y., Chang, C.H.: An improved autonomous cross-layer optimization framework for wireless multimedia communication. In: 2014 IEEE/ACIS 13th International Conference on Computer and Information Science (ICIS), pp. 53–58, June 2014
21. El-atty, S.M.A.: Efficient packet scheduling with pre-defined QOS using cross-layer technique in wireless networks. In: 11th IEEE Symposium on Computers and Communications (ISCC 2006), pp. 820–826, June 2006
22. Essa, A.A., Zhang, X., Wu, P., Abuzneid, A.: ZigBee network using low power techniques and modified LEACH protocol. In: 2017 IEEE Long Island Systems, Applications and Technology Conference (LISAT), pp. 1–5, May 2017

23. Fröhlich, A.A., Okazaki, A.M., Steiner, R.V., Oliveira, P., Martina, J.E.: A cross-layer approach to trustfulness in the Internet of Things. In: 16th IEEE International Symposium on Object/Component/Service-Oriented Real-Time Distributed Computing (ISORC 2013), pp. 1–8, June 2013

24. Fu, F., van der Schaar, M.: A new systematic framework for autonomous cross-layer optimization. IEEE Trans. Veh. Technol. **58**(4), 1887–1903 (2009)

25. Gomez, C., Paradells, J., Bormann, C., Crowcroft, J.: From 6LoWPAN to 6Lo: expanding the universe of IPv6-supported technologies for the Internet of Things. IEEE Commun. Mag. **55**(12), 148–155 (2017)

26. Granjal, J., Monteiro, E., Silva, J.S.: Security for the Internet of Things: a survey of existing protocols and open research issues. IEEE Commun. Surv. Tutor. **17**(3), 1294–1312 (2015)

27. Gutierrez, J.A., Naeve, M., Callaway, E., Bourgeois, M., Mitter, V., Heile, B.: IEEE 802.15.4: a developing standard for low-power low-cost wireless personal area networks. IEEE Netw. **15**(5), 12–19 (2001)

28. Han, C., Jornet, J.M., Fadel, E., Akyildiz, I.F.: A cross-layer communication module for the Internet of Things. Comput. Netw. **57**(3), 622–633 (2013)

29. Hansen, C.J.: Internetworking with Bluetooth low energy. GetMobile Mob. Comput. Commun. **19**(2), 34–38 (2015)

30. Hasan, N., Ali, M., Barradas, A., Correia, N.: Cross-layer optimization for reliability improvement of data delivery in 6LoWPAN-based networks. In: 2015 14th Annual Mediterranean Ad Hoc Networking Workshop (MED-HOC-NET), pp. 1–7, June 2015

31. Hong, Y.G., Gomez, C., Sangi, A.R., Aanstoot, T.: IPv6 over Constrained Node Networks (6Lo) Applicability & Use cases. Internet-Draft draft-ietf-6lo-use-cases-01, Internet Engineering Task Force. Work in Progress. https://datatracker.ietf.org/doc/html/draft-ietf-6lo-use-cases-01

32. Hu, P.: A system architecture for software-defined industrial Internet of Things. In: 2015 IEEE International Conference on Ubiquitous Wireless Broadband (ICUWB), pp. 1–5, October 2015

33. Huang, P., Xiao, L., Soltani, S., Mutka, M.W., Xi, N.: The evolution of MAC protocols in wireless sensor networks: a survey. IEEE Commun. Surv. Tutor. **15**(1), 101–120 (2013)

34. Huber, P.J.: The behavior of maximum likelihood estimates under nonstandard conditions. In: Proceedings of the Fifth Berkeley Symposium on Mathematical Statistics and Probability, Volume 1: Statistics, pp. 221–233. University of California Press, Berkeley (1967)

35. Iqbal, M., Naeem, M., Anpalagan, A., Ahmed, A., Azam, M.: Wireless sensor network optimization: multi-objective paradigm. Sensors **15**(7), 17572–17620 (2015)

36. Ishaq, I., et al.: IETF standardization in the field of the Internet of Things (IoT): a survey. J. Sens. Actuator Netw. **2**(2), 235–287 (2013)

37. Kafi, M.A., Othman, J.B., Badache, N.: A survey on reliability protocols in wireless sensor networks. ACM Comput. Surv. **50**(2), 31:1–31:47 (2017)

38. Kellerer, H., Pferschy, U., Pisinger, D.: Knapsack Problems. Springer, Heidelberg (2004). https://doi.org/10.1007/978-3-540-24777-7

39. Kim, H.S., Ko, J., Culler, D.E., Paek, J.: Challenging the IPv6 routing protocol for low-power and lossy networks (RPL): a survey. IEEE Commun. Surv. Tutor. **19**(4), 2502–2525 (2017)

40. Kumar, K., Kumar, S., Kaiwartya, O., Cao, Y., Lloret, J., Aslam, N.: Cross-layer energy optimization for IoT environments: technical advances and opportunities. Energies **10**(12) (2017)

41. Latchman, H.A., Katar, S., Yonge, L., Gavette, S.: HomePlug AV and IEEE 1901: A Handbook for PLC Designers and Users, 1st edn. Wiley-IEEE Press, Hoboken (2013)
42. Le, N.T., Jang, Y.M.: Energy-efficient coverage guarantees scheduling and routing strategy for wireless sensor networks. Int. J. Distrib. Sen. Netw. **11**(8), 612383 (2015)
43. Levis, P., Patel, N., Culler, D., Shenker, S.: Trickle: a self-regulating algorithm for code propagation and maintenance in wireless sensor networks. In: Proceedings of the First USENIX/ACM Symposium on Networked Systems Design and Implementation (NSDI), pp. 15–28 (2004)
44. Liu, Y., Seet, B.C., Al-Anbuky, A.: Ambient intelligence context-based cross-layer design in wireless sensor networks. Sensors **14**(10), 19057–19085 (2014)
45. Marais, J.M., Malekian, R., Abu-Mahfouz, A.M.: LoRa and LoRaWAN testbeds: a review. In: 2017 IEEE AFRICON, pp. 1496–1501, September 2017
46. Marler, R., Arora, J.: Survey of multi-objective optimization methods for engineering. Struct. Multidiscip. Optim. **26**(6), 369–395 (2004)
47. Mathur, S., Saha, D., Raychaudhuri, D.: Cross-layer MAC/PHY protocol to support IoT traffic in 5G: poster. In: Proceedings of the 22nd Annual International Conference on Mobile Computing and Networking, MobiCom 2016, pp. 467–468. ACM, New York (2016)
48. Mikhaylov, K., Petaejaejaervi, J., Haenninen, T.: Analysis of capacity and scalability of the LoRa low power wide area network technology. In: European Wireless 2016; 22th European Wireless Conference, pp. 1–6, May 2016
49. Mirjalili, S., Mirjalili, S.M., Lewis, A.: Grey wolf optimizer. Adv. Eng. Softw. **69**, 46–61 (2014)
50. Nguyen, H.X., Roughan, M.: Rigorous statistical analysis of internet loss measurements. IEEE/ACM Trans. Netw. **21**(3), 734–745 (2013)
51. Park, M.: IEEE 802.11ah: sub-1-GHz license-exempt operation for the Internet of Things. IEEE Commun. Mag. **53**(9), 145–151 (2015)
52. Patil, M., Biradar, R.C.: A survey on routing protocols in wireless sensor networks. In: 2012 18th IEEE International Conference on Networks (ICON), pp. 86–91, December 2012
53. Pompili, D., Akyildiz, I.F.: A multimedia cross-layer protocol for underwater acoustic sensor networks. IEEE Trans. Wirel. Commun. **9**(9), 2924–2933 (2010)
54. IROL Power, Lossy Networks (ROLL): IETF routing over low power and lossy networks (ROLL), January 2016. http://datatracker.ietf.org/doc/charter-ietf-roll/
55. Ratasuk, R., Vejlgaard, B., Mangalvedhe, N., Ghosh, A.: NB-IoT system for M2M communication. In: 2016 IEEE Wireless Communications and Networking Conference, pp. 1–5, April 2016
56. Ray, P.P., Agarwal, S.: Bluetooth 5 and Internet of Things: potential and architecture. In: 2016 International Conference on Signal Processing, Communication, Power and Embedded System (SCOPES), pp. 1461–1465, October 2016
57. Ray, P.: A survey on Internet of Things architectures. J. King Saud Univ. Comput. Inf. Sci. **30**, 291–319 (2016)
58. Resner, D., de Araujo, G.M., Fröhlich, A.A.: On the impact of dynamic routing metrics on a geographic protocol for WSNs. In: 2016 VI Brazilian Symposium on Computing Systems Engineering (SBESC), pp. 109–115, November 2016
59. Resner, D., de Araujo, G.M., Fröhlich, A.A.: Design and implementation of a cross-layer IoT protocol. Sci. Comput. Program. (2017)

60. Roh, H.T., Lee, J.W.: Cross-layer optimization for wireless sensor networks with RF energy transfer. In: 2014 International Conference on Information and Communication Technology Convergence (ICTC), pp. 919–923, October 2014
61. Shrestha, B.: Analysis of hybrid CSMA/CA-TDMA channel access schemes with application to wireless sensor networks. Ph.D. thesis, The University of Manitoba, Winnipeg, July 2013. Hybrid CSMA
62. Siddavaatam, P., Sedaghat, R., Sharma, A.T.: intel-LEACH: an optimal framework for node selection using dynamic clustering for wireless sensor networks. In: 2017 12th IEEE International Conference for Internet Technology and Secured Transactions (ICITST), pp. 136–146, December 2017
63. Siddavaatam, P., Sedaghat, R., Sharma, A.T.: A novel multi-objective optimization approach for design flow in high level synthesis, vol. 55, pp. 990–1004 (2018)
64. Tanenbaum, A., Wetherall, D.: Computer Networks. Pearson Prentice Hall, Upper Saddle River (2011)
65. Thubert, P., Wetterwald, P., Vasseur, J.P., Michel, E.: Reverse directed acyclic graph for multiple path reachability from origin to identified destination via multiple target devices (2015)
66. Vilajosana, X., Wang, Q., Chraim, F., Watteyne, T., Chang, T., Pister, K.S.J.: A realistic energy consumption model for TSCH networks. IEEE Sens. J. **14**(2), 482–489 (2014)
67. Vuran, M.C., Akyildiz, I.F.: Error control in wireless sensor networks: a cross layer analysis. IEEE/ACM Trans. Netw. **17**(4), 1186–1199 (2009)
68. Ye, W., Heidemann, J., Estrin, D.: Medium access control with coordinated adaptive sleeping for wireless sensor networks. IEEE/ACM Trans. Netw. **12**(3), 493–506 (2004)

Design and Analysis of Energy Efficient Wireless Body Area Network (WBAN) for Health Monitoring

Aakriti Khanna[✉], Vaibhav Chaudhary, and Sindhu Hak Gupta

Amity School of Engineering and Technology, Amity University, Noida, India
aakritikhanna.8@gmail.com, c5.vaibhav.1995@gmail.com, shak@amity.edu

Abstract. Wireless Body Area Network (WBAN) is an upcoming research area of Wireless Sensor Networks. In WBAN, wireless sensors are either implanted within human body or worn over body to examine physiological strictures like blood pressure rate, body temperature, blood glucose level. Main emphasis is laid on making WBAN energy efficient. Network stability and increased throughput are also areas of prime concern for WBAN. Modified SIMPLE (Stable Increased Multi-hop Protocol Link Efficiency) cluster level protocol (M-SIMPLE) has been proposed and implemented. M-SIMPLE is capable of performing single and multi-hop whereas in most of the other existing protocols either of the one is possible. A cost function is required to select a parent node which comprises maximum residual energy and has minimum distance from sink. Optimal number of cluster heads are chosen to increase the efficiency. Improved stability and packet delivery is achieved which further improves the efficiency of the network.

1 Introduction

1.1 Brief Description

With rapid growth and advancements of technology, people are opting for wireless health care systems. Recent discoveries in wireless networking, microelectronics and wireless sensor help to cope with eminent problems and social challenges of ageing population. This population and rising cost of health care have triggered the introduction of novel technology which avail us the opportunity of early detection and prevention from well-known diseases. The goal is to reduce the efforts of humans and provide early treatment.

This well-structured health care system consists of miniature devices that can either be implanted within human body or worn over easily. This contributes to Wireless Body Area Network (WBAN). Such devices monitors physiological conditions non-instructively and improves the functionality of the system. As stated by IEEE task group 6, WBAN is defined as "a low power device operating in or around human body serving applications including medical, personal

© Springer-Verlag GmbH Germany, part of Springer Nature 2018
M. L. Gavrilova and C. J. K. Tan (Eds.): Trans. on Comput. Sci. XXXIII,
LNCS 10990, pp. 25–39, 2018.
https://doi.org/10.1007/978-3-662-58039-4_2

entertainment and other" [4]. WSN has been optimised in different sectors like agriculture, medicine [5], and multimedia. The primary architecture of WBAN supports 3-tier system. In 1st tier, physiological parameters are measured using sensors. Followed by 2nd tier which is the handling device or sink where all the data is collected and which is further transmitted to 3rd tier which is server station which stores all the related data.

The comparable studies proposes that with the help of WBAN elderly people, who find it difficult to go to hospital for regular check-ups can be monitored at home. WBAN in comparison to Wireless Sensor Networks (WSN) faces certain challenges like Network Topology, High data rates and Power Supply being the most common. The major drawback of Wireless Body Area Network is that sensor nodes have limited energy which further results in early discharge of sensors. In WSN a lot of energy efficient routing protocols are implemented. But the feasibility of these protocols is difficult in accordance to that of WBAN applications. WBAN supports low power consumption and dynamic device support. It can also be inferred that, WBAN provides better data transmission in reference to individual use.

The overall network power consumption depends primarily on the power consumed by transceivers and the parting power required for performing operations of components like switching. In order to achieve energy efficiency [9], the transceivers adjust their transmitting power in accordance to their destination. This is either accomplished by accounting plenty of small hops (i.e. multi-hop) or limited number of larger hops (i.e. single-hop). Single hopping is more efficient in handling network parameters like end-to-end relay, data packet loss etc. whereas multi hopping maintains power efficiency. Besides using these individually, we can plot together the advantages of both single and multi-hop in attaining the two approaches of energy efficient wireless network such that, computable path loss between transmitter and receiver and also power consumption. Even in practical scenario the combinational hopping is considered to be effective (Fig. 1).

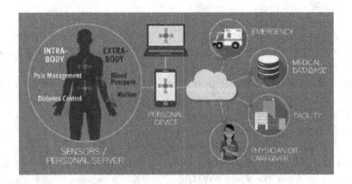

Fig. 1. Architecture of Wireless Body Area Network (WBAN)

1.2   Background

WBAN is dependent on a variety of physiological sensors for user applications. These sensors generate analog signals which are computed and modified according to capabilities of wireless network. These nodes regularly transmit status of events and keeps end user informed at the time of emergency situations.

The design issues faced by WBAN as mentioned [17] are power consumption of sensors, communication range and characteristics of sensor for transmission. System configuration and user interface also prove as a challenge in designing.

Our main focus is towards making WBAN energy efficient by maintaining power consumption of sensors. The crucial factor for end user application is to make minimise the size of device and make it light weighted. The on-sensors have the calibre to save power and extend battery life.

While considering a system design [15] incremental cooperative communication for WBANs [14] has been analysed. The author proposed three-stage incremental relay. This 3-node relaying was compared with single and two-stage relaying schemes. Simulation outcomes depicted that the three-stage relaying scheme had higher throughput with reduced Packet Error Rate (PER). The first approach was to make WBAN more appropriate for routing which lead to improvement in transmission rate. In [10] Threshold sensitive Energy Efficient sensor Network (TEEN) protocol was simulated for sending the data of patient in WBAN and different critical events were monitored for its demonstration. Simulation results proved TEEN to be a suitable protocol to control data transmission of a patient's body. It showed more efficiency in terms of critical events and time, making it more beneficial for WBAN system. Different types of routing schemes are followed for transmission and reception of data. In [9], comparison of single-hop and multi-hop was analysed. Results prove that single-hop transmission is more efficient when consumption of power of wireless sensor node's transceivers are considered.

In [16], a cluster based routing protocol for WBANs, named as Energy Efficient Adaptive Routing in WBAN (EAR-BAN) was proposed. It enables Body Nodes (BNs) to form clusters and the co-ordinator (BNC) elects cluster heads. The operational model of this protocol consists of four phases-synchronization phase, cluster set-up phase, gateway selection phase and centralized cluster formation phase. It performed centralized operation which reduced the computational burden of body nodes. This was an important attempt towards reduction of power consumption and led to power saving as the main power source becomes cluster head.

The main contribution of this work is:

1. A stable routing protocol has been proposed. It has maximum residual energy and leads to high throughput. Comparative analysis of combination of single and multi-hop protocol with only multi-hop protocol is made.
2. Optimal Number of cluster heads are chosen to make proposed system energy efficient.

3. nRF24XXX is implemented and it is observed that this sensor along with proposed M-SIMPLE protocol decreases the dead nodes by a considerable percentage.
4. Proposed protocol increases the throughput by huge margin.
5. Path Loss and Residual Energy are also improved for the proposed system.

Further, the article is organised as follows: Model designing is discussed in Sect. 2, in Sect. 3 the proposed algorithm is analysed. Section 4 computes simulation results which proves the proposed protocol is energy efficient and in Sect. 5 we conclude.

2   Model Designing

2.1   Energy Model

The sensors in WBAN are available with limited resources, so different routing schemes are applied to make transmission favourable and maintain energy consumption. The node deployment is done such that it reduces link failures and results in low power dissipation and data loss. The aim is to design such model which satisfies following needs such that

(a) It maintains higher stability which can reduce the overall energy consumption of nodes leading it to be available for longer time.
(b) Also, low consumption contributes to higher throughput.

The assumptions made to calculate energy consumption is that the analysing and the processing energy are negligible with respect to overall energy of the proposed network. Therefore, total energy consumption is represented by equation as used in [1]

$$E_{tot} = E_{Tx} - E_{Rx} \quad \text{for all wireless nodes.} \tag{1}$$

The energy dissipated by transmitter and receiver are E_{Txelec} and E_{Rxelec}. Also amplifier circuit energy is denoted by E_{amp} and distance between two nodes are denoted by D_{ij}. The transmission energy is computed from the equation:

$$E_{Tx} = w[T_{xelec} + E_{amp}(n_{ij})D_{ij}^{n_{ij}}] \tag{2}$$

and reception energy is computed from relation:

$$E_{Rx} = w[E_{Rxelec}] \tag{3}$$

where w is total number of bits received or transmitted (Figs. 2 and 3).

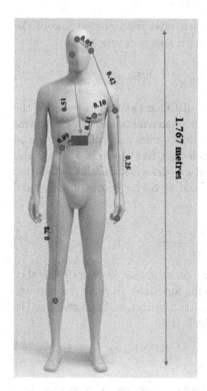

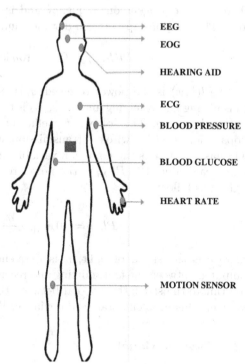

Fig. 2. Network establishment of proposed protocol

Fig. 3. Deployment of sensors on human body

2.2   Model Computing Path Loss

Path loss is stated as the reduction in power density and its SI units is decibels (dB). Additive White Gaussian Noise (AWGN) is used to reduce the signal attenuation. Power is radiated outwards by the transmitting antenna and anything between the transmitting and the receiving antennas causes the signal to deteriorate. In WBAN, body movement variations affects the transmitted signal. Path loss equation is dependent on frequency and distance as expressed in [12].

$$Pl(h, x) = Pl(h) \times Pl(x). \tag{4}$$

where, $Pl(h)$ represents path loss due to frequency and $Pl(x)$ represents path loss due to distance.

The relationship between frequency and path loss is shown as

$$\sqrt{Pl(h)} \propto h^m. \tag{5}$$

where m is contingent on frequency and is associated with the outline of the body. The relationship between distance and path loss is expressed as

$$Pl(h, x) = Pl_O + 10n \log_{10} \frac{x}{xo} + Y\sigma. \tag{6}$$

where $Pl(h, x)$ is the power received, x is the distance calculated between the transmitting and receiving antennas, xo is the reference distance, The coefficient of path loss is n. The following coefficient value varies according to type of propagation used for wireless transmission such that free space has $n = 2$ and LOS has $n = 4$. Y is the Gaussian variable and σ represents standard deviation from mean value [13]. Pl_{OR} is the power received at reference distance xo and is given as follows

$$Pl_{OR} = 10 \log_{10} \frac{(4\pi \times x \times h)^2}{c}. \tag{7}$$

where c is the speed of light, h is frequency of signal, x is overall distance computed between the transmitting and receiving antennas. In reality, it is tough to estimate the strength of signal between the transmitter and the receiver. To overcome this problem, standard deviation $Y\sigma$ is used.

2.3   Network Model

In this work, the biosensors are connected via set of nodes called relays for data collection. These act like backbone of transmission network. While designing the network the main approach is towards position of relays as biosensors and sink have fixed positions.

As stated in [3], a wireless link is established within a communication range Rc as well as sensors having range Rs where $Rc > Rs$.

The sensor coverage area parameter is given as:

$$a_{ij} = \begin{cases} 1 & \text{if sensor can establish link with relay} \\ 0 & \text{elsewhere} \end{cases}$$

and sink coverage parameter is given as:

$$e_{jk} = \begin{cases} 1 & \text{if relay is establish with sink} \\ 0 & \text{elsewhere} \end{cases}$$

The connectivity between two sites or points is given as:

$$b_{jl} = \begin{cases} 1 & \text{if } j \text{ and } l \text{ are connected with link} \\ 0 & \text{elsewhere} \end{cases}$$

where i belongs to S which represents set of sensors, j belongs to set of sites and k belongs to set of sinks.

The next hop node is selected from scheme of cost function as stated in [1]. The cost function at every node decides forwarder node or parent node. It is computed from the following expression:

$$CF(i) = \frac{x(i)}{R(i)}$$

where, i = computes the number of nodes
$x(i)$ = represents the distance between the sensor node and the sink, and
$R(i)$ = represents computed value of residual energy

The node which has minimum cost function is elected as forwarder which collects the data and further transmits to sink.

3 Proposed Algorithm

The proposed algorithm is designed such that it supports all the network parameters such as area of application (Human Body) and the number of devices involved. When all parameters are set, this adds to completion of initialization and set up phase. When the transmission phase begins, the collective data from nodes is sent to destination node. It is expected that receiving node has enriched energy source and it's assumed that it does not have any limitation of energy. The flow of information in the proposed protocol is represented by Fig. 4.

The algorithm opts for most reasonable outlook depending upon sensor positioning and overall residual energy of every node. The algorithm is based on two phases:

3.1 Setup Phase

Nodes are active at the position where they have been deployed. The communication count defines node data that is forwarded. A threshold limit is setup for communication.

3.2 Operational Phase

As all the nodes are deployed at their fixed position in beginning. Among them one of the node is selected as destination node which is in proximity of transmission.

In multi-hop protocol data from all source nodes is transmitted to destination node via multiple hops. The next hop node is selected from neighbors' list which is on range proximity.

This may be done in two ways:
Method 1: distance of next hop between node and destination should be lower than that of selected node of destination and also energy level should be greater than threshold value.
Method 2: Method 1 is put into action when other parameters like remaining energy, distance to be covered etc. are considered.

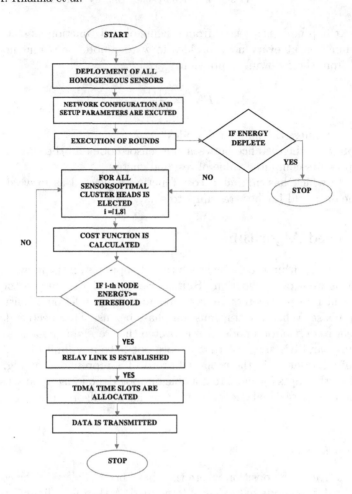

Fig. 4. Flowchart of proposed method

But in order to make it more efficient we have chosen combination of single and multi-hop protocol. In addition to multi-hop, single hop aims to nodes within coverage range of reception node whereas, to transmit outside range we use multi-hop.

For transmission of data nRF24L01 has been used. It belongs to Nordic family of series nRF24LXX. It is highly integrated, ultra low power (ULP) 2 Mbps RF transceiver. It is a single chip transceiver with frequency ranging from 2.4–2.48 GHz. The front end of radio model uses Gaussian frequency shift Keying (GFSK) modulation. It has higher data rate available with dual power saving modes which makes it more suitable for low power models. The values of transceiver are mentioned in Table 1.

It works on the principle of First in First out (FIFO) which ensures smooth data flow between radio model and microcontroller unit.

Table 1. nRF24L01 transceiver parameters

Parameters	nRF 24L01
Tx DC Current	11.3 mA
Rx DC Current	12.3 mA
Minimum Supply Voltage	1.90 V
Etx-elec	18 nJ/b
Erx-elec	20 nJ/b
Eamp	2e−7 J/b

4 Simulation Results

4.1 Network Life Time

The protocol's stability is shown in Fig. 5. This protocol used a combination of single and multi-hop. Cost function played a significant role in maintaining the consumption of energy of the nodes. It was used to select a new parent node or forwarder in every round. Figure 5 clearly depicts that M-SIMPLE protocol had better stable time period. This occurred due to electing appropriate new parent node in every round. Therefore, each node utilises equal energy in every round and also all the nodes die within same time limit. Whereas in the protocol which used only multi-hop, the parent nodes selected alternate longer routes as the temperature increased which lead to more energy consumption. Therefore, the nodes die early (Table 2).

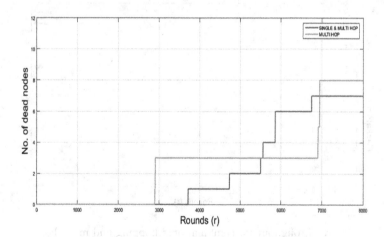

Fig. 5. Comparison of combinational hopping with multi-hop over network

Table 2. Optimized values for maintaining network lifetime

Rounds (X)	Number of dead nodes (Y) single and multi-hop	Number of dead nodes (Y) multi-hop	Approximate efficiency %
3000	0	3	1
4000	1	3	66.66
5000	2	3	50
6000	6	3	50
7000	7	8	14.28
8000	7	8	14.28

4.2 Throughput

Throughput states as the total number of successful packets that are received at the sink. Since WBAN has collective data of patients, therefore a protocol is required which maximises successful data transmission and minimises packet drop. As depicted M-SIMPLE protocol (single and multi-hop) achieves better throughput as compared to ATTEMPT protocol (multi-hop). Number of successful packets depends on the number of alive nodes. Greater number of alive nodes means more successful packet reception which will in turn increase the overall throughput of the network. The number of successful packets are decreased in ATTEMPT due to less stability as compared to M-SIMPLE (Fig. 6 and Table 3).

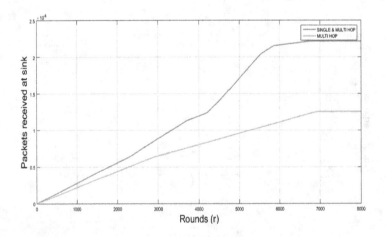

Fig. 6. Throughput for combinational hopping and multi-hop

Table 3. Optimized values for maintaining throughput

Rounds (X)	Packets received at sink (Y) single and multi-hop	Packets received at sink (Y) multi-hop	Approximate efficiency %
1000	0.2806	0.2166	6.4
2000	0.5577	0.4342	22.08
3000	0.8890	0.6427	27.70
4000	1.196	0.80	33.11
5000	1.70	0.9510	44.05
6000	2.163	1.109	48.72
7000	2.21	1.25	43.43
8000	2.21	1.25	43.43

4.3   Residual Energy

The energy expenditure in each round of a network is known as residual energy which is presented in the above figure. The nodes which are far from the sink use a parent node to transfer their data to the sink. This parent node is elected with the help of aforementioned cost function. Energy is saved in each round by selecting an appropriate parent node. For the transmission of packets to the sink, the proposed protocol used a new parent node in each round which reduced the overloading of any particular node. Results after simulation showed that the combinational protocol consumed minimum energy. Whereas in multi-hop topology, some of the nodes were exhausted early due to heavy traffic load (Fig. 7 and Table 4).

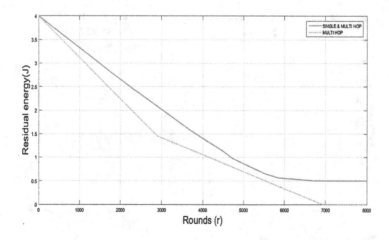

Fig. 7. Residual energy of combinational hopping and multi-hop

Table 4. Optimized values for maintaining overall residual energy

Rounds (X)	Residual energy (Y) single and multi-hop	Residual energy (Y) multi-hop	Approximate efficiency %
0	4	4	0
1000	3.319	3.317	0.06
2000	2.655	2.248	15.32
3000	2.017	1.412	29.99
4000	1.4	1.05	25
5000	0.862	0.692	19.72
6000	0.554	0.334	22
7000	0.5	0	100
8000	0.5	0	100

4.4 Path Loss

The final figure represents the system's path loss. It is defined as a function
of distance and frequency. The path loss calculated in Fig. 8 is a function of
distance. The simulation results showed that M-SIMPLE protocol (single and
multi-hop) reduced the path loss. This was because the transmission due to
multi-hop topology decreases the distance and results in minimum path loss. The
proposed protocol performs good initially. But after 3000 rounds, path loss of
ATTEMPT (multi-hop) decreases at a high rate. This is because in this protocol
some nodes die early. Therefore, less number of active nodes meant less path loss.
Since M-SIMPLE protocol had more active nodes and longer stability period,
this meant that the path loss will be more (Table 5).

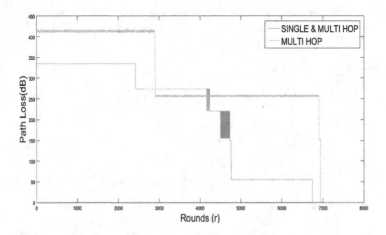

Fig. 8. Path loss of combinational hopping and multi-hop

Table 5. Optimized values for comparison over pathloss

Rounds (X)	Path loss (Y) single and multi-hop	Path loss (Y) Multi-hop	Approximate efficiency %
0	335.2	416.7	24.31
1000	335.2	415.9	24.07
2000	335.2	412.5	23.06
3000	273.4	256.9	6.29
4000	273.8	256.6	6.28
5000	55.06	258.9	78.73
6000	55.06	255.5	78.45
7000	0	0	0
8000	0	0	0

Overall observed results are further concluded in the following Table 6:

Table 6. Overall evaluation of proposed protocol

Parameters	m-Simple	Attempt	Approx. Efficiency %
Number of dead nodes	7	8	12.5
Throughput	2.21	1.25	96
Residual Energy	0.5	0	100
Path Loss	55.06	255.5	78.2

From the following table we inferred that M-SIMPLE protocol is more efficient in comparison to ATTEMPT protocol.

5 Conclusion

In this work, energy efficient protocol for routing data in WBANs has been proposed. The combination of single & multi-hop transmission in comparison with only multi-hop is implemented and analyzed. A new forwarder or parent node has been elected in each round for transmission which has been decided by cost function. The simulation results showed that the proposed protocol enhanced packet delivery to sink and the network stability time by 70 % and 35 % respectively. Further improvements may be made by implementing cooperative signaling for heterogeneous body area network. Results of the current work seem very encouraging and are capable of making WBAN energy efficient.

References

1. Nadeem, Q., Javaid, N., Mohammad, S.N., Khan, M.Y., Sarfraz, S., Gull, M.: SIM-PLE: stable increased-throughput multi-hop protocol for link efficiency in wireless body area networks, broadband and wireless computing. In: 2013 Eighth International Conference Communication and Applications (BWCCA), pp. 221–226 (2013)
2. Sharma, R., Ryait, H.S., Gupta, A.K.: Performance analysis of ATTEMPT, SIM-PLE and DEEC routing protocols in WBAN. Int. J. Latest Trend Eng. Technol. (IJLTET) 6(2), 133–139 (2015)
3. Elias, J., Mehaoua, A.: Energy-aware topology design for wireless body area networks. In: IEEE ICC 2012 - Selected Areas in Communications Symposium, pp. 3409–3410 (2012)
4. IEEE Standard for Local and Metropolitan Area Networks-Part 15.6: Wireless Body Area Networks, IEEE Standard 802.15.6-2012,802.15 Working Group, February 2012
5. Ahmad, J., Zafar, F.: Review of body area network technology & wireless medical monitoring. Int. J. Inf. Commun. Technol. Res. 2(2), 186–188 (2012)
6. Liu, B., Yan, Z., Chen, C.W.: Medium access control for wireless body area network with QoS provising and energy efficient design. IEEE Trans. Mob. Comput. 16(2), 422–434 (2017)
7. Kim, Y., Lee, S.S., Lee, S.K.: Coexistence of ZigBee-based WBAN and WiFi for health telemonitoring system. IEEE J. Biomed. Health Inform. 20(1), 222–230 (2016)
8. Chunqiang, H., Hongjuan, L., Yan, H., Tao, X., Xiaofeng, L.: Secure and efficient data communication protocol for wireless body area networks. IEEE Trans. Multi-Scale Comput. Syst. 2(2), 94–107 (2016)
9. Pešović, U.M., Mohorko, J.J., Benkič, K., Čučej, Ž.F.: Single-hop vs. Multi-hop - Energy efficiency analysis in wireless sensor networks. In: 18 Telekomunikacioni forum TELFOR, pp. 471–474 (2010)
10. Gupta, S., Kaur, P.: WBAN health monitoring system using TEEN protocol: threshold sensitive energy efficient network protocol. IJISET Int. J. Innov. Sci. Eng. Technol. 2(10), 20–25 (2015)
11. Jasawat, U., Pandey, N.: Adaptive network coding mechanism for network lifetime extension based on body area network. Int. J. Innov. Res. Comput. Commun. Eng. 4(6), 11588–11594 (2016)
12. Javaid, N., Khan, N.A., Shakir, M., Khan, M.A., Bouk, S.H., Khan, Z.A.: Ubiquitous healthcare in wireless body area networks - a survey. J. Basic Appl. Sci. Res. 3(4), 747–759 (2013)
13. Rappaport, T.S.: Wireless Communications: Principles and Practice, vol. 2. Prentice Hall PTR, New Jersey (1996)
14. Huang, X., Shan, H., Shen, X.: On energy efficiency of cooperative communications in wireless body area network. In: IEEE Wireless Communications and Networking Conference, pp. 1097–1101 (2011)
15. Yousaf, S., Javaid, N., Qasim, U., Alrajeh, N., Khan, Z.A., Ahmed, M.: Towards reliable and energy-efficient incremental cooperative communication for wireless body area networks. Artic. Sens. 16, 284 (2016)
16. ul Huque, Md.T.I., Munasinghe, K.S., Abolhasan, M., Jamalipour, A.: EAR-BAN: energy efficient adaptive routing in Wireless Body Area Networks. In: 7th International Conference on Signal Processing and Communication Systems (ICSPCS), pp. 1–10 (2013)

17. Jovanov, E., Milenkovic, A., Otto, C., de Groen, P.C.: A wireless body area network of intelligent motion sensors for computer assisted physical rehabilitation. PMC-US National Library of Medicine National Institutes of Health, p. 6 (2005)
18. Gupta, S.H., Singh, R.K., Sharan, S.N.: Performance analysis of coded cooperation and space time cooperation with multiple relays in Nakagami-m fading. In: Gavrilova, M.L., Tan, C.J.K., Saeed, K., Chaki, N., Shaikh, S.H. (eds.) Transactions on Computational Science XXV. LNCS, vol. 9030, pp. 172–185. Springer, Heidelberg (2015). https://doi.org/10.1007/978-3-662-47074-9_10
19. Sethi, D., Bhattacharya, P.P.: A study on Energy Efficient and Reliable Data Transfer (EERDT) protocol for WBAN. In: Second International Conference on Computational Intelligence & Communication Technology (CICT), pp. 254–258 (2016)
20. Smail, O., Kerrar, A., Zetili, Y., Cousin, B.: ESR: energy aware and stable routing protocol for WBAN networks. In: International Wireless Communications and Mobile Computing Conference (IWCMC), pp. 452–457 (2016)
21. Promwongsa, N., Sanguankotchakorn, T.: Packet size optimization for energy-efficient 2-hop in multipath fading for WBAN. In: 22nd Asia-Pacific Conference on Communications (APCC), pp. 445–450 (2016)
22. Verma, M., Rai, R.: Energy-efficient cluster-based mechanism for WBAN communications for healthcare applications. Int. J. Comput. Appl. **120**(19), 24–31 (2015)
23. Sindhu, S., Vashisth, S., Chakarvarti, S.K.: A review on Wireless Body Area Network (WBAN) for Health monitoring system: implementation protocols. In: Communications on Applied Electronics (CAE), vol. 4, no. 7, pp. 16–20. Foundation of Computer Science FCS, New York (2016). ISSN 2394-4714
24. Singh, S., Negi, S., Uniyal, A., Verma, S.K.: Modified new-attempt routing protocol for wireless body area network. In: International Conference on Advances in Computing, Communication, & Automation (ICACCA) (Fall), September 2016, pp. 1–5 (2016)
25. Javaid, N., Ahmad, A., Nadeem, Q., Imran, M., Haider, N.: iM-SIMPLE: iMproved stable increased-throughput multi-hop link efficient routing protocol for Wireless Body Area Networks. Comput. Hum. Behav. **51**, 1003 (2014). Article in Press
26. Javaid, N., Abbas, Z., Fareed, M.S., Khan, Z.A., Alrajeh, N., M-ATTEMPT: a new energy-efficient routing protocol for wireless body area sensor networks. In: The 4th International Conference on Ambient Systems, Networks and Technologies (ANT 2013), pp. 224–231 (2013)

A Revamp Approach for Training of HMM to Accelerate Classification of 16S rRNA Gene Sequences

Prakash Choudhary[1]([✉]) and M. P. Kurhekar[2]

[1] Department of Computer Science and Engineering,
National Institute of Technology Manipur, Imphal 795001, India
choudharyprakash@nitmanipur.ac.in
[2] Department of Computer Science and Engineering,
Visvesvaraya National Institute of Technology Nagpur, Nagpur 440010, India

Abstract. In the era of Information Technology, the field of Bioinformatics is rapidly growing with research in various related topics. The database of biological information has become much higher than its consumption. Automatic classification of biological information is one of the critical problems in Bioinformatics. Therefore, the critical issue is to regulate and manage the enormous amount of novel information to facilitate access to this useful and valuable biological information. The specific nucleus dilemma in classifying biological information is the annotation of various biological sequences with functional features. Annotation of the significant and rapidly increasing amount of genomic sequence data requires computational tools for classification of genes in DNA sequences. This paper presents a computational method for classification of highly conserved 16S rRNA biological sequences. We took Biological sequence classification as motivation to reveal a methodology that uses Hidden Markov Models (HMMs) to classify them. This paper explains the description of the algorithms used for implementing three phases of HMM (training, decoding, and evaluation) to classify sequences into clusters that have known similar functional properties. In the implementation of the training phase, we have addressed practical issues like initial parameter selection for HMM and computational weakness for the large data set. Later in the paper, we have shown that methodology presents a classification accuracy of 91% for Bacillus and 97% for Clostridia.

Keywords: HMM · Parameters estimation · Biological sequence
16S rRNA · Gene classification · Bioinformatics

1 Introduction

The DNA sequence string represents the alphabet with the letters A, C, G and T standing for the nucleotides or bases adenine, cytosine, guanine and thymine.

© Springer-Verlag GmbH Germany, part of Springer Nature 2018
M. L. Gavrilova and C. J. K. Tan (Eds.): Trans. on Comput. Sci. XXXIII,
LNCS 10990, pp. 40–60, 2018.
https://doi.org/10.1007/978-3-662-58039-4_3

In the recent era, the DNA sequences of many organisms have been ascertained. Discovering the genes is crucial and usually a prerequisite for moreover analysis such as the characterization of the function of the gene product, determining the phylogeny of different species or understanding gene direction. The challenge for bioinformatics projects is that only accessible information comes in the form of biological sequence data (RNA, DNA, and protein sequences). The other kind of above details or domain knowledge is considered to be inexistent [1].

The problem of classifying genes in a genomic DNA sequence is challenging and despite the extensive effort that has not been solved yet satisfactorily. The accuracy of gene classification and foresight algorithms should be high adequate to furnish reliable information. The results of such algorithms are used for the automated annotation of genomes, and there is a demand for dynamic gene finders which are as accurate as possible. The research area of finding and classification of genes sequence by computational methods has become very competitive. The classification of the biological sequence into functional and structural families based on sequence homology is a principal problem in computational biology.

Classification of gene sequences has emerged as an essential tool to diagnose infections in clinical laboratories. For example, in many cases of diseases, it is not possible to identify the bacteria or the virus immediately that has caused the disease. Therefore, it becomes difficult to start treatment for these types of unknown bacteria or virus. After identification of specific bacteria, one can begin to proper treatment. Thus, it would be useful to identify the bacteria family and start first practice till further diagnostic tests are awaited. This paper deals with an approach of automatically classifying bacteria into their respective families.

In our approach, we have select two bacteria families who have their properties and biological structures, Bacillus and Clostridia of gene 16S rRNA sequences [2]. Bacillus and Clostridia are highly conserved gene across different species. The gene target that is most commonly used for bacterial identification is 16S rRNA [3] (or 16S rDNA), an ~1500 base pair gene that codes for a portion of the 30S ribosome. The 16S rRNA gene sequencing has emerged as a more accurate and faster method to identify a wide variety of aerobic and anaerobic bacteria and used widely in clinical laboratories.

To assist in discovering the patterns among sets of these sequences, we use the well-developed theory of Hidden Markov Models (HMMs). HMMs have been successfully used in speech recognition systems for clustering different looking sound patterns of same speech part or phoneme [4]. The hidden Markov model (HMM) is a statistical model used in many fields including bioinformatics, econometrics and population genetics [5]. The problem of classification of biological sequences is very similar to clustering different looking sequences for the same class of genome. However, the difference is like information to be processed by the HMMs. In the biological sequence domain, the probability density functions are defined over a discrete domain - the realm of the alphabet with which the sequences are formed.

The sequence classification methods can be divided into three large categories [6] - feature based classification [7–11], distance-based classification [12–15] and model-based classification [16–21].

In the training process, A k-order Markov model applied to classify protein and text sequence data [24]. Instead of the conventional generative setting the model is trained in a discriminative setting to increase the classification power of the generative model-based methods. Different from Markov Model, Hidden Markov Model (HMM) assumes that the system being modeled is a Markov process with unobserved states. The authors explained uses of HMMs as well as all the aspects of HMMs [23]. They even show results of their model on several different machines. This work is also mainly theory-related work, and they fail to mention issues related to implementation such as training, and initialization. They discuss several problems but not enough details on implementation issues. The work in [26] uses HMM to identify the recurrent short structural 3D building blocks (SBBs) taking into account the local dependence between them. In [27], discusses multiple sequence alignments but does not discuss the implementation details. Also, [28] focuses more on the biological aspect of this process rather than the implementation aspect. [25] use a profile HMM to classify sequences. In [37] authors present a mini review about the 16S rRNA gene sequencing for bacterial identification in the diagnostic laboratory: Pluses, Perils, and Pitfalls. A study of the taxonomic classification of bacterial 16S rRNA genes using short sequencing reads presents [40]. Most of the work we examined, gives mathematical details as well as results but no details on the training phase.

Our objective is to train and use HMMs for clustering the biological sequences. We have implemented a system for HMMs, including all three phases (training, decoding, and evaluation) of an HMM. We trained each dataset individually with HMM to represent them as one cluster. Once the HMM models have been generated and trained, we compared an unidentified sequence with these models to determine the one which has the closest match to the sequence. This comparison gave us a likelihood probability which measures the likelihood that the sequence belongs to the class represented by the HMM. The primary advantage of HMMs over, distance-based sequence matching algorithms is that the HMMs can automatically estimate, or be trained for, a cluster of unaligned sequences.

Our exploratory study has shown some promise and has uncovered some potential problems with the approach. The model, while being trained, converged to a very good representation of the training set with the significant success rate. The rest of the paper is organized as: Sect. 2 presents the implementation prerequisite details. In Sect. 3 we have addressed the training phase challenges. Section 4 presented the experimental results followed by discussion and conclusion Sect. 5.

2   Preliminaries and Method

A Markov technique is an uncertain version, in which the state belongs to a finite set, the evolution happens at a discrete time and the probability distribution of a state at a given time is explicitly structured simplest on the preceding state [33]. Hidden Markov model is one of the most machine learning models in information extraction and retrieval. Hidden Markov version (HMM) is a sort of a statistical model wherein the machine being modeled is assumed to be a Markov technique with unknown parameters, and the venture is to determine the hidden parameters, from the observable parameters, based totally in this assumption. Thus, its underlying statistics and its parameters have been adjusted in such a way that it describes the system of its study in the best way [34].

Figure 1 shows the flowchart to classify Biological sequence on the HMM process. The first step in this process is to initialization of starting parameters for HMM (calculating the initial state probabilities, the transition probabilities, and the emission probabilities). Once these values have been initialized, we uses an algorithm Baum-Welch (BW) [22] based on two auxiliary recursive procedures called forward and backward algorithms [5]. The next step is learning of model - requires the Baum-Welch Algorithm where an iterative approach is applied on the given observation sequence to maximize likelihood for given observation sequence until the model converges and a feasible model is obtained. Once a feasible model is generated, then we proceed to evaluation phase of HMM. Evaluation phase of HMM uses the Forward and the Backward algorithm to test and validate the learned model.

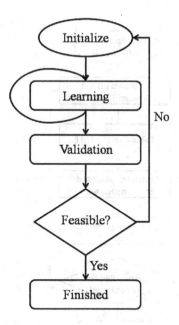

Fig. 1. Flow chart for the HMM process

- Initial State Probability(π_i) Initial State Probability π_i is an array of length N (number of states), which stores the initial probability of being at state s.

- Emission Probability($B(o_t)$) - B is a two-dimensional array ($M \times N$), with B[M][N] representing the probability of generating output symbol "o" at time "t" from state s. M is the number of unique symbols (in our datasets (gene 16S rRNA) 5 symbol present).

$$b_j(o_t) = P(o_t|q_t = j), \quad 1 \le t \le T(length\ of\ observation\ sequences) \quad (1)$$

- Transition Probability(A_{ij}) - A is a two-dimensional array ($N \times N$) representing the probability of going to state j from state i .

$$a_{ij} = P(q_t = j \mid q_{t-1} = i), \qquad 1 \le i,\ j \le N \qquad (2)$$

Figure 2 shows the flowchart for the learning part of our model. The learning starts with an initial parameter. In learning phase of HMM forward and backward algorithm apply for each observation sequence. Once each of the biological sequences have been passed through forward, backward algorithms, value of $\xi_t(i, j)$ (expected number of transitions from state i, to state j) and $\gamma_t(j)$ (expected number of time state j emit a symbol) are calculated. On the basic of forward, backward, $\xi_t(i, j)$ and $\gamma_t(j)$, value of emission, transition and initial state probabilities have been updated. We repeat this process for total number of steps which is predefined or until the converged is reached.

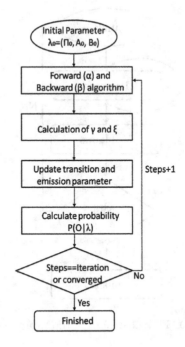

Fig. 2. Flow chart for the learning of model

3   Challenges in Training Phase

In the implementation of the training phase, we have addressed practical issues like initial parameter selection for HMM and computational weakness for large data set. During the training phase of HMM, starting parameter selection decides how quickly the results converge. Also, while applying the Baum-Welch algorithm, we have reused the parameters which have converged in less number of iterations during training of the HMMs.

3.1   Starting Parameter Selection

Training of HMM is very sensitive to initial parameters selection and appropriate selection also useful to avoid local maxima problem. Three initialization strategies, namely, count-based, random and uniform can be used to set initial parameter for HMM [29]. All of these methods used to set initial parameter for the training of HMM, increase the possibility of inaccuracy in trained HMM, because in these methods the possibility of getting stuck observation probability into local maxima is higher.

The assumptions made by us in setting the initial parameters for HMM are as given below:

- The HMM model used in our work is a fully-connected HMM which allows revisiting previous states, meaning that the hidden states are fully-connected.
- The discrete observations densities are taken as datasets.
- States are assumed to be persistent ($P(S_t = i|S_{t+1} = i)$ is relatively high), instead of flipping frequently.
- The numbers of states is assumed to be known.

With this proposed method, the learned parameters can be used as initial parameters for Baum-Welch learning algorithm to avoid the faults caused by starting with random parameters.

In our work, we used an approximation-based HMM learning method to estimate initial parameter for HMM. Approximation-based HMM learning has a better likelihood $P(O|\lambda)$ and it converged in less number of iterations as compared to the randomly initialized method, and it also avoids the problem of getting stuck into local maxima.

As our work based on above assumption we set initial parameter $\lambda_{0=}(\pi_0, A_0, B_0)$ for HMM having $P(S_t = i|S_{t+1} = i)$ is relatively high around 0.95 to 1 and randomly distribute remaining probability among all others parameters of transition probability distribution. In initial state distribution, we equally distribute probability among all-state as a starting state probability. By these assumptions, we train HMM for the predefined number of iteration or until convergence reached. We can estimate the probability of initial parameters for HMM by calculating simple counting. Let S_i represents the current state, S_j represents the next state. Let N represent the set of all states and R represent remaining probability after assign $P(S_t = i|S_{t+1} = i)$. The symbol #(*) represents the number of *. The initial parameters can be calculated as below:

- Initial state distribution:

$$\pi = \left[\begin{array}{ccccc} \frac{\#(S=1)}{\#(N)} & \frac{\#(S=2)}{\#(N)} & \frac{\#(S=3)}{\#(N)} & \cdots & \frac{\#(S=N)}{\#(N)} \end{array} \right]_{1 \times N}$$

- Transition probability distribution:

$$A = \left[\begin{array}{cccc} 0.95 \; to \; 1 & \frac{R}{\#(N-1)} & \cdots & \frac{R}{\#(N-1)} \\ \frac{R}{\#(N-1)} & 0.95 \; to \; 1 & \cdots & \frac{R}{\#(N-1)} \\ \vdots & \vdots & \ddots & \vdots \\ \frac{R}{\#(N-1)} & \frac{R}{\#(N-1)} & \cdots & 0.95 \; to \; 1 \end{array} \right]_{N \times N}$$

- Observation probability distribution: Observation probabilities are set using a pseudo-random number generator. The observation matrix is guaranteed to have non-zero probabilities for any allowable observations. The observation probabilities for any given state always sum to 1.

3.2 Parameter Reuse of Baum-Welch Algorithm

HMM models learns from training data using the Baum-Welch algorithm. It's an iterative process for estimating HMM parameters and guaranteed to reach converged, but have some drawback such as local maxima and convergence requires more number of iterations.

Starting parameter selection and the issue of the number of iterations can be reduced in the following manner: Usually the $O(t)$ length of observation sequences is in high dimensions such as in our work Bacillus and Clostridia have large number of string. The Baum-Welch algorithm is trained incrementally on all observation sequences considering one by one observation sequence, at each value of t it consider input (o_t) and computes probability of this observation sequence. Here we consider a common parameter $S_0(x_0)$ on all observation sequence $O = (2, 3, \ldots, T-1)$, which stores the maximum probability for this observation symbol calculated during training. And this probability used during the time of probability calculation for next observation sequence (o_{t+1}). So the ratio of observation probability for next iteration is as follows:

$$\frac{b_j(o)}{b_0(x)} = \frac{b_j(o_j)}{b_0(x_j)} \; , \qquad 1 \leq j \; \leq N \qquad (3)$$

The left hand side of equation is the likelihood function of observation probability scaled by factor $b_0(x)$. The right-hand side of the equation gives the testing criteria for the sufficiency of x_j for state S_j vs. state S_0. The dimensions at each state j are sufficient to discriminate it from the common state S_0. For time steps $t = 2, 3, \ldots ., T$, all the parameter estimation procedure is same except for the state likelihood function $b_j(x)$.

The model $\lambda = (\pi, A, B)$ has three terms to describe namely the state transition probability distribution A, the initial state distribution and the observation symbol probability distribution B. Next, for updating the model parameters we need a couple of auxiliary variables $\gamma_t(i)$ and $\xi_t(i,j)$, and their calculations only depend on $\alpha_t(i)$ and $\beta_t(i)$, to describe the procedure for re-estimation.

The re-estimation procedure for the HMM parameters using parameter reuse is as [31].

The Re-estimation of HMM Parameters

A. In Forward Procedure

1. **Initialization:**

$$\alpha_1(j) = \pi_j \frac{b_j(o_j[1])}{b_0(x_0[1])}, \qquad (1 \le j \le N) \qquad (4)$$

2. **Induction:**
$for(1 \le t \le T-1, 1 \le j \le N)$

$$\alpha_{t+1}(j) = [\sum_{j=1}^{N} \alpha_j(i)a_{ij}] \frac{b_j(o_j)[t+1]}{b_0(x_0)[t+1]} \qquad (5)$$

3. **Termination:**

$$\frac{p(x([1],\ldots\ldots,x[t])|\lambda)}{p(x([1],\ldots\ldots,x[t])|S_0)} = \sum_{i=1}^{N} \alpha_T(i) \qquad (6)$$

Where S_0 is the condition that state S_0 is true at every t.

B. In Backward Procedure

1. **Initialization:**

$$\beta_t(i) = 1 \qquad (7)$$

2. **Induction:** $for(t = T-1, \ldots\ldots, 1, 1 \le i \le N)$

$$\beta_t(j) = [\sum_{i=1}^{N} a_{ji}] \frac{b_i(o_i)[t+1]}{b_0(x_0)[t+1]} \beta_{t+1}(i) \qquad (8)$$

C. Re-estimation in BW Training Phase:

1. Re-estimation method of (γ) is same as estimated by the conventional BW.
2. Re-estimation of (ξ)

$$\xi_t(i,j) = \frac{\alpha_t(i)a_{ij} \frac{b_j(o_j)[t+1]}{b_0(x_0)[t+1]} \beta_{t+1}(j)}{\sum_{i=1}^{N} \sum_{m=1}^{N} \alpha_t(i) a_{im} \frac{b_m(o_m)[t+1]}{b_0(x_0)[t+1]} \beta_{t+1}(m)} \qquad (9)$$

We described the reuse of parameters for the training of Hidden Markov Models that use maximum probability of observation sequence in the forward and backward procedures. The HMMs trained in this way can reduce the number of iteration for convergence.

3.3   Parallel Training of Baum-Welch Algorithm

Baum-Welch algorithm is a robust way for estimation of parameters in HMM. It guarantees to converge, but dealing with the large observation sequence it is slow and usually requires a lot of memory. The algorithm execution time can be improved by parallel execution of the forward algorithm and backward algorithm, which calculates probabilities that uses Baum-Welch algorithm during the training phase to iteratively update the parameter λ until the difference of data likelihood between two iterations is small. In our approach, we have the large length of observation sequence so to normalize parameters within a range we have used scaling factor. A scaling factor is calculated in the forward procedure, which is used in both forward and backward algorithms to normalize.

The algorithm execution time can be improved by parallel execution of the forward algorithm and backward algorithm, which calculates probabilities that uses Baum-Welch algorithm during the training phase to iteratively update the parameter λ until the difference of data likelihood between two iterations is small. During each iteration the algorithm calculates the following values:

$$\alpha_t(i) = P(o_1...o_t, q_t = s_i | \lambda), \quad i = 1, 2, ...N \tag{10}$$

$$\beta_t(i) = P(o_{t+1}...o_T | q_t = s_i, \lambda) \quad i = 1, 2, ...N \tag{11}$$

The forward probability value $\alpha_t(i)$ is calculated from the first variable (q_1) to the last (q_t), while The backward probability $\beta t(i)$ is calculated from the last variable (q_t) back to the first (q_1) reverse of forward procedure.

As shown in Fig. 3 we have divide observation sequence (O) in blocks. Each processor then runs the parallel version of Baum-Welch algorithm for each block [32]. For example, processor1 will deal with $q_{1,1}, q_{1,2}, ..., q_{1,s}$ (s is number of states). Two things are then needed to be solved: How to communicate between adjacent blocks when calculating $\alpha_t(i)$ and $\beta_t(i)$, and how do the processors work together to update the model parameters. In the example, the length of observation sequence is 6 and its run on 2 processors; each processor dealing with

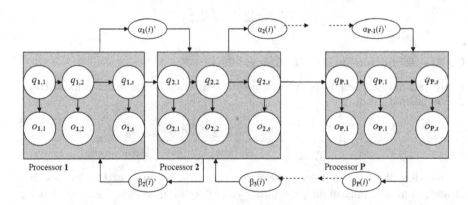

Fig. 3. Parallel training of Baum-Welch algorithm

one block (3 observation sequence). At each iteration, sequential Baum-Welch algorithm needs to first calculate $\alpha_t(i)$ from q_1 to q_6, then $\beta_t(i)$ back from q_6 to q_1. After both $\alpha_t(i)$ and $\beta_t(i)$ are available, it then uses them to update the model parameter λ.

For the parallel Baum-Welch algorithm, the two processors can simultaneously calculate its own $\alpha_{p,t}(i)$ and $\beta_{p,t}(i)$ (p is the index of processor). However, Processor1 cannot wait for the $\beta_{2,0}$ to start the backward calculation, while Processor2 cannot wait for $\alpha_{1,3}$ for its forward calculation. So instead, we use the value of such $\alpha_{p,t}(i)$ and $\beta_{p,t}(i)$ values of their previous iteration instead.

In parallelized training of Baum-Welch, we need to update $\lambda = \{A, B, \pi\}$ at each iteration. π is trivial so can just let on the first processor to update it; For A and B, we will use a'_{ij} and $b_i(o_k)'$ respectively, where $\gamma_{p,t}(i)$ and $\xi_{p,t}(i,j)$ are the respective local parameters within the pth processor:

$$a'_{ij} = \frac{\sum_{p=1}^{P} \sum_{t=1}^{T-1} \xi_{p,t}(i,j)}{\sum_{p=1}^{P} \sum_{j=0}^{N-1} \sum_{t=1}^{T-1} \xi_{p,t}(i,j)} \tag{12}$$

$$b_i(o_k)' = \frac{\sum_{p=1}^{P} \sum_{t=1, o_t=o_k}^{T} \gamma_{p,t}(i)}{\sum_{p=1}^{P} \sum_{t=1}^{T} \gamma_{p,t}(i)} \tag{13}$$

The experiment performs on the $Intel Xeon$ $E5440$ 2.83 GHz CPU, equipped with 16 GB RAM. We used the platform openMP with GCC compiler to run the parallel algorithm and achieved a speedup 7.20x on eight processors.

3.4   Scaling of Parameter to Avoid Floating-Point Underflow

During the training of HMM, long observation sequences often result in the computation of extremely small probabilities. These values are usually lower in magnitude than the smallest value a normal floating point number a system can define. This result in a significant problem called floating-point underflow. This numerical instability is seen in the Forward, the Backward and the Baum-Welch Algorithms during the training phase of HMM. When these algorithms are applied to long sequences, it results in extremely small probability values that could underflow on most machines [30]. We solve this problem differently for each algorithm.

3.4.1   Scaling the Forward and Backward Variables

The calculation of $\alpha_t(i)$ and $\beta_t(i)$ involves multiplication with probabilities. All these probabilities have a value less than 1 (generally significantly less than 1), and as t starts to grow large, each term of $\alpha_t(i)$ or $\beta_t(i)$ starts to head exponentially to zero. For sufficiently large t (e.g., 100 or more) the dynamic range of $\alpha_t(i)$ and $\beta_t(i)$ computation will exceed the precision range of essentially any machine (even in double precision). The basic scaling procedure multiplies $\alpha_t(i)$ by a scaling coefficient that is dependent only of the time t and independent of the state i. The scaling factor for the forward variable is denoted c_t (scaling is done every time t for all states $i - 1 \leq i \leq N$). This factor will also be used for scaling the backward variable, $\beta_t(i)$. Scaling $\alpha_t(i)$ and $\beta^t(i)$ with the same scale factor will show useful in parameter estimation.

Consider the computation of the forward variable, $\alpha_t(i)$. In the scaled variant of the forward algorithm some extra notations will be used. $\alpha_t(i)$ denote the unscaled forward variable, $\widehat{\alpha}_t(i)$ denote the scaled and iterated variant of $\alpha_t(i)$, $\widehat{\widehat{\alpha}}_t(i)$ denote the local version of $\alpha_t(i)$ before scaling and c_t will represent the scaling coefficient at each time. Here follows the scaled forward algorithm:

1. Initialization

$$\text{Set } t = 2;$$

$$\alpha_1(i) = \pi_i b_i(o_1), \quad 1 \leq i \leq N \tag{14}$$

$$\widehat{\alpha}_t(i) = \alpha_1(i), \quad 1 \leq i \leq N \tag{15}$$

$$c_1 = \frac{1}{\sum_{i=1}^{N} \widehat{\alpha}_t(i)} \tag{16}$$

$$\widehat{\alpha}_1(i) = c_1 \alpha_1(i) \tag{17}$$

2. Induction

$$\widehat{\widehat{\alpha}}_t(i) = b_i(o_t) \sum_{j=1}^{N} \widehat{\alpha}_{t-1}(j) a_{ji}, \quad 1 \leq i \leq N \tag{18}$$

$$c_t = \frac{1}{\sum_{i=1}^{N} \widehat{\widehat{\alpha}}_{t_t}(i)} \tag{19}$$

$$\widehat{\alpha}_t(i) = c_t \widehat{\widehat{\alpha}}_t(i), \quad 1 \leq i \leq N \tag{20}$$

3. Update time

Set $t = t + 1$;
Return to step 2 if $t \leq T$;
Otherwise, terminate the algorithm (goto step 4).

4. Termination

$$P(O|\lambda) = \frac{1}{\prod_{t=1}^{T} c_t} \quad or \quad logP(O|\lambda) = -\sum_{t=1}^{T} logc_t \tag{21}$$

The main difference between the scaled and the non-scaled forward algorithm lies in steps two and four. In step two, (20) can be rewritten if (18) and (19) are used:

$$\widehat{\alpha}_t(i) = c_t \widehat{\widehat{\alpha}}_t(i)$$

$$= \frac{1}{\sum_{k=1}^{N} \left(b_k(o_t) \sum_{j=1}^{N} \widehat{\alpha}_{t-1}(j) a_{jk} \right)} \cdot \left(b_i(o_t) \sum_{j=1}^{N} \widehat{\alpha}_{t-1}(j) a_{ji} \right) \tag{22}$$

$$= \frac{b_i(o_t) \sum_{j=1}^{N} \widehat{\alpha}_{t-1}(j) a_{ji}}{\sum_{k=1}^{N} \left(b_k(o_t) \sum_{j=1}^{N} \widehat{\alpha}_{t-1}(j) a_{jk} \right)}, \quad 1 \leq i \leq N \tag{23}$$

By induction, the scaled forward variable can be found in terms of the none scaled as:

$$\widehat{\alpha}_{t-1}(j) = \left(\prod_{r=1}^{t-1} c_T\right) \alpha_{t-1}(j), \quad 1 \le j \le N \tag{24}$$

The ordinary induction step can be found as (same as (4.6) but with one time unit shift):

$$\alpha_t(i) = b_i(o_t) \sum_{j=1}^{N} \alpha_{t-1}(j) a_{ji}, \quad 1 \le i \le N \tag{25}$$

With (24) and (25) it is now possible to rewrite (22) as:

$$
\begin{aligned}
\widehat{\alpha}_t(i) &= \frac{b_i(o_t) \sum_{j=1}^{N} \widehat{\alpha}_{t-1}(j) a_{ji}}{\sum_{k=1}^{N} b_k(o_t) \sum_{j=1}^{N} \widehat{\alpha}_{t-1}(j) a_{jk}} \\[2mm]
&= \frac{b_i(o_t) \sum_{j=1}^{N} \left(\prod_{r=1}^{t-1} c_r\right) \alpha_{t-1}(j) a_{ji}}{\sum_{k=1}^{N} b_k(o_t) \sum_{j=1}^{N} \left(\prod_{r=1}^{t-1} c_r\right) \alpha_{t-1}(j) a_{jk}} \\[2mm]
&= \frac{\left(\prod_{r=1}^{t-1} c_r\right) \left(b_i(o_t) \sum_{j=1}^{N} \alpha_{t-1}(j) a_{ji}\right)}{\left(\prod_{r=1}^{t-1} c_r\right) \sum_{k=1}^{N} \left(\left(b_k(o_t) \sum_{j=1}^{N} \alpha_{t-1}(j) a_{jk}\right)\right)} \\[2mm]
&= \frac{\alpha_t(i)}{\sum_{k=1}^{N} \alpha_t(k)}, \quad 1 \le i \le N
\end{aligned}
\tag{26}
$$

As (26) shows, each $\alpha_t(i)$ is scaled by the sum over all states of $\alpha_t(i)$ when the scaled forward algorithm is applied.

The termination (step 4) of the scaled forward algorithm, evaluation of $P(O|\lambda)$, must be done in a different way, because the sum of $\widehat{\alpha}_T(i)$ cannot be used, it is scaled already. However the following properties can be used:

$$\prod_{r=1}^{T} c_r \sum_{i=1}^{N} \alpha_T(i) = 1 \tag{27}$$

$$\prod_{r=1}^{T} c_r * P(O|\lambda) = 1 \tag{28}$$

$$P(O|\lambda) = \frac{1}{\prod_{r=1}^{T} c_r} \tag{29}$$

As (29) shows can $P(O|\lambda)$ be found, but the problem is that if (29) is used the result will still be very small (and probable out of the dynamic range for a computer). If the logarithm is taken on both sides the following equation can be used:

$$\log P(O|\lambda) = \frac{1}{\prod_{r=1}^{T} c_r} = -\sum_{t=1}^{T} \log c_t \tag{30}$$

This is exactly what is done in the termination step of the scaled forward algorithm. The logarithm of $P(O|\lambda)$ is often just as useful as $P(O|\lambda)$, because in most cases, this measure is used as a comparison with other probabilities (for other models).

The scaled backward algorithm can be found more easily, since it will use the same scale factor as the forward algorithm. The notations used is similar to the forward variables notations, $\beta_t(i)$ denote the unscaled backward variable, $\widehat{\beta}_t(i)$ denote the scaled and iterated variant of $\beta_t(i)$, $\widehat{\widehat{\beta}}_t(i)$ denote the local version of $\beta_t(i)$ before scaling and c_t will represent the scaling coefficient at each time. Here follows the scaled backward algorithm:

1. Initialization

$$Set, \quad t = T - 1; \tag{31}$$

$$\beta_T(i) = 1, \ 1 \leq i \leq N \tag{32}$$

$$\widehat{\beta}_T(i) = c_T\beta_T(i), \quad 1 \leq i \leq N \tag{33}$$

2. Induction

$$\widehat{\widehat{\beta}}_t(i) = \sum_{j=1}^{N} \widehat{\beta}_{t+1}(j)\, a_{ij}b_j\left(o_{t+1}\right), \quad 1 \leq i \leq N \tag{34}$$

$$\widehat{\beta}_t(i) = c_t\widehat{\widehat{\beta}}_t(i), \quad 1 \leq i \leq N \tag{35}$$

3. Update time

Set $t = t - 1$;
Return to step 2 if $t > 0$;
Otherwise, terminate the algorithm.

3.4.2 Scaling of Training Variables During Baum-Welch Algorithm

With scaled variables $\xi_t(i, j)$ will be as:

$$
\begin{aligned}
\xi(i,j) &= \frac{\widehat{\alpha}_t(i)\, a_{ij}b_j\left(o_{t+1}\right) \widehat{\beta}_{t+1}(j)}{\sum_{i=1}^{N}\sum_{j=1}^{N} \widehat{\alpha}_t(i)\, a_{ij}b_j\left(o_{t+1}\right)\widehat{\beta}_{t+1}(j)} \\[2mm]
&= \frac{\left(\prod_{r=1}^{t} c_r\right)\alpha_t(i)\, a_{ij}b_j\left(o_{t+1}\right)\left(\prod_{r=t+1}^{T} c_r\right)\beta_{t+1}(j)}{\sum_{i=1}^{N}\sum_{j=1}^{N}\left(\prod_{r=1}^{t} c_r\right)\alpha_t(i)\, a_{ij}b_j\left(o_{t+1}\right)\left(\prod_{r=t+1}^{T} c_r\right)\beta_{t+1}(j)} \\[2mm]
&= \frac{\left(\prod_{r=1}^{t} c_r\right)\left(\prod_{r=t+1}^{T} c_r\right)\alpha_t(i)\, a_{ij}b_j\left(o_{t+1}\right)\beta_{t+1}(j)}{\left(\prod_{r=1}^{t} c_r\right)\left(\prod_{r=t+1}^{T} c_r\right)\sum_{i=1}^{N}\sum_{j=1}^{N}\alpha_t(i)\, a_{ij}b_j\left(o_{t+1}\right)\beta_{t+1}(j)} \\[2mm]
&= \frac{\alpha_t(i)\, a_{ij}b_j\left(o_{t+1}\right)\beta_{t+1}(j)}{\sum_{i=1}^{N}\sum_{j=1}^{N}\alpha_t(i)\, a_{ij}b_j\left(o_{t+1}\right)\beta_{t+1}(j)} \tag{36}
\end{aligned}
$$

$\gamma_t(i)$ are the same if scaled or not scaled such as:

$$
\begin{aligned}
\gamma_t(i) &= \frac{\widehat{\alpha}_t(i)\,\widehat{\beta}_t(j)}{\sum_{i=1}^{N}\widehat{\alpha}_t(i)\,\widehat{\beta}_t(i)} \\
&= \frac{\left(\prod_{r=1}^{t} c_r\right)\alpha_t(i)\left(\prod_{r=t+1}^{T} c_r\right)\beta_t(i)}{\sum_{i=1}^{N}\left(\prod_{r=1}^{t} c_r\right)\alpha_t(i)\left(\prod_{r=t+1}^{T} c_r\right)\beta_t(i)} \\
&= \frac{\left(\prod_{r=1}^{t} c_r\right)\alpha_t(i)\left(\prod_{r=t}^{T} c_r\right)\alpha_t(i)\,\beta_t(i)}{\left(\prod_{r=1}^{t} c_r\right)\alpha_t(i)\left(\prod_{r=t}^{T} c_r\right)\sum_{i=1}^{N}\alpha_t(i)\,\beta_t(i)} \\
&= \frac{\alpha_t(i)\,\beta_t(i)}{\sum_{i=1}^{N}\alpha_t(i)\,\beta_t(i)} \qquad\qquad (37)
\end{aligned}
$$

As (36) and (37) shows, $\xi_t(i,j)$ and $\gamma_t(i)$ are the same if scaled or not scaled. Since π, A and B uses $\xi_t(i,j)$ and $\gamma_t(i)$ for calculation, these probabilities will also be independent of which forward or backward variables are used (scaled or unscaled). Thereby -

$$
\pi_i = \frac{\widehat{\alpha}_t(i)\,\widehat{\beta}_t(j)}{\sum_{i=1}^{N}\widehat{\alpha}_T(i)} \qquad\qquad (38)
$$

$$
a_{ij} = \frac{\sum_{t=1}^{T-1}\widehat{\alpha}_t(i)\,a_{ij}b_j(o_{t+1})\,\widehat{\beta}_{t+1}(j)}{\sum_{t=1}^{T-1}\widehat{\alpha}_t(i)\,\widehat{\beta}_t(j)} \qquad\qquad (39)
$$

$$
b_j(o_k) = \frac{\sum_{t=1,o_t=o_k}^{T}\gamma_t(j)}{\sum_{t=1}^{T}\gamma_t(j)} \qquad\qquad (40)
$$

All these algorithms mentioned in above section solve the problem of clustering biological sequences into different groups according to their functionality or sequence structure. The simulation of HMM-based on these revamp approaches for training of HMM is presented in the following section of the paper, to construct a model for each class of biological sequence.

4   Experimental Results and Analysis

4.1   Training and Observation of HMM for Bacillus Database

Bacillus database is the first dataset used for our simulation. It contains the gene sequence of 940 different species belong with bacillus family. After training of HMM, we have generated a model which is a specific model for the Bacillus dataset. The result of this simulation allows us to understand the capabilities of our model. Based on the results of this dataset, we can classify the given biological sequence into Bacillus or non Bacillus.

Figure 4 shows traces of probability of observation of given sequences $P(O \mid \lambda)$ as the increment in a number of iteration for the training of HMM with

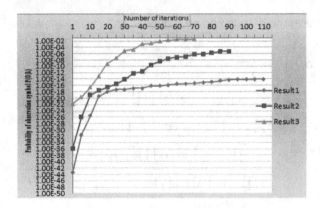

Fig. 4. Training of HMM for Bacillus dataset (Color figure online)

Table 1. Confusion matrix for the validation of trained HMM model for Bacillus dataset

Dataset type	Bacillus	Non-Bacillus
Bacillus	91%	9%
Clostridia	2%	98%

improvement in the given algorithm. Result1 (Blue line) in this graph shows the probability of observation sequence with the traditional Baum-Welch (BW) algorithm. Result2 (Red line) shows the result of modified BW algorithm. As compared to simple BW algorithm, it gives better results as probability $P(O \mid \lambda)$ of observation sequence is higher at each iteration and it converged around 85 iterations while traditional BW converged around 115 iterations. Result3 (Green line) shows the results obtained by implementing BW algorithm with appropriate starting parameters. We can observe that in Result3, observation probability $P(O \mid \lambda)$ converged in very less number of iterations around 65.

We have also experimented with another dataset of Clostridia (for validation of our model with unknown datasets) classification among Bacillus and Non-bacillus on our given trained HMM. We have tried to validate our model using given datasets of known types. Table 1 shows results of our experimentation on trained HMM for different datasets like Bacillus and Clostridia. We can observe from these results that accuracy of trained HMM is around 91% for given Bacillus type dataset. Also in the Non-Bacillus dataset like clostridia, it is showing the accuracy of about 98%.

4.2 Training and Observation of HMM for Clostridia Database

In this Gene, 16S rRNA Clostridia is considered a dataset for our simulation of the sequences. It contains total 449 different species DNA sequence. After training of HMM, we have generated a model which is a specific model for the

Clostridia dataset. The result of this simulation allows us to understand the capabilities of our model. Based on the results of this dataset, we can classify the given biological sequence into Clostridia or non-Clostridia. Figure 5 shows traces of probability of observation $P(O \mid \lambda)$ as the increment in number of iteration for the training of HMM with improvement in the given algorithm. Result1 (Blue line) in this graph shows the probability of observation sequence with the simple Baum-Welch algorithm. Result2 (Red line) shows the result of modified BW algorithm. As compared to simple BW algorithm, it gives better results, probability $P(O \mid \lambda)$ of observation sequence is higher at each iteration, and it converged around 75 iterations while traditional BW converged around 95 iterations. And the Result3 (Green line) shows the results obtained by implementing BW algorithm with appropriate starting parameter. It shows the best result obtained as compared to other two. We can observe that in Result3, observation probability $P(O \mid \lambda)$ converged in very less number of iterations around 45 and thus improve the efficiency of the algorithm with accuracy. This approach also avoids getting trapped in local maxima. We have taken input files of two datasets named Bacillus and Clostridia, which are in the same format as explained in the previous section. We have tried to validate our model using given datasets of known types. Table 2 shows results of our experimentation on trained HMM for different datasets like Bacillus and Clostridia. We can observe from these results that accuracy of trained HMM is around 97% for given Clostridia type dataset. Also in Non-Clostridia dataset like Bacillus, it is showing the accuracy of about 94%.

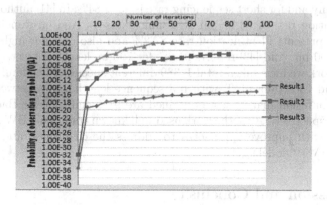

Fig. 5. Training of HMM for Clostridia dataset (Color figure online)

4.3 Comparative Analysis on 16S rRNA Identification

To the best of our knowledge the actual taxonomic composition of the microbial communities residing in different environments is limited, evaluating the classification accuracy and specificity of any taxonomic classification algorithm on 16S rRNA data sets obtained from actual environmental samples is difficult.

Table 2. Confusion matrix for the validation of trained HMM model for Clostridia dataset

Dataset type	Clostridia	Non-Clostridia
Clostridia	97%	3%
Bacillus	6%	94%

The comparison is difficult for the 16S rRNA gene sequence classification performed on different environmental (uncultured) bacteria and nucleotide length (bp) sequence. As expected, we found that performance varied widely across different regions, sequencing strategies, and ranks.

A Nave Bayes classifier has applied on the simulated data set consists of 1677 sequences of lengths around 800. The classifier were obtained a range of confidence score thresholds ranging between 75% and 100% [35]. The performance of C16S was evaluated using a synthetically created 16S rDNA data set available at the FAMeS (Fidelity of Analysis of Metagenomic Samples) repository [36]. Under this simulated scenario, assignments of the sequences in the FAMeS dataset were obtained using C16S and obtained average confidence score thresholds around 90%.

In a study by [38], have found that 16S yielded species identification rates of 62% to 91%. And for the bacteria closely related the identification rates were lower with (62% to 83%) than the values traditionally acceptable in the clinical laboratory such as 90%) [39]. The identification rate achieved in [40] is 70%–85% for gene family and for short sequencing rate is 63%–84%. In [41] authors reports an identification rate of 83.1% on GenBank datasets. And on same GenBank dataset [42] presents identification rate 87.2% and also consider high sequence length of 1500 bp. And [43] reports 91.6% for 796 bp length dataset.

In our work we have taken total 1, 389 different species of two families Bacillus (940) and Clostridia (449) with full sequence length varies between 700 to 1700. The confusion matrix score obtained are 91% and 97% for Bacillus and Clostridia respectively. Also, the work used well known traditional BW algorithm for training with improve approaches to address the challenges often occur in training of HMM model.

5   Discussion and Conclusion

In this work, we have described an alignment-free classification methodology that uses Hidden Markov Models for classification of 16S rRNA bacterial families Bacillus and Clostridia. The main focus of this work is on training phase of HMM which uses well known Baum-Welch algorithm with improved approach for learning of biological sequence and an efficient computation to re-estimation of the parameter after each iteration. In the training of HMM having two main tasks, the first is the selection of an appropriate starting parameter for HMM and the second is an estimation of transition and emission probabilities for the model. We aimed to develop a model for each cluster of biological sequences in which each sequence consists of the different species name. From observation probability, we can classify sequences of which clusters they belong.

During implementation, we have addressed the weaknesses related to HMM training phase. Initial parameter selection is a major issue for HMM training phase because previously parameters are set according to random, count and unit based method. None of this technique has given *best fit for all* solution for classification. In each learning phase, we need to set new random parameters hence this model provides different observed probability $P(O \mid \lambda)$ based on the initial parameters. Because of this, HMM training was trapped in local maxima. Hence, right parameters selection at time of learning HMM resolves this local maxima problem. We have shown that the proposed method to select right starting parameters, speeds up the learning of the HMM models with fewer iterations and better likelihood values. Better likelihood means better accuracy and faster results.

We have found large computation probability during training phase which requires long execution time for large datasets. We have reduced execution time during the training phase of HMM by the parallelizing Baum-Welch algorithm. Thus our results are more efficient and optimized. When implementing an HMM for long observation sequences, often it results in the computation of extremely small valued probabilities. These values are usually smaller in magnitude than the smallest value a normal floating point number a system can define. This result in a significant problem called floating-point underflow. In our case, we have consider total 1389 species gene sequence with length about 700 to 1700 bp. Hence, at the time of training due to a large number of observation sequences, computation goes out of range for the machine. So we have added iterative scaling to overcome this problem.

Our results obtained from the simulation of HMM, indicate that the HMMs learn the pattern embedded in the sequences of the training set. In all our tests we construct two models, one for bacillus second for clostridia. While analyzing the results we found that accuracy of HMM for bacillus is around 91% and the accuracy of HMM for clostridia is around 97%.

References

1. Ferles, C., Beaufort, W.-S., Ferle, V.: Self-Organizing Hidden Markov Model Map (SOHMMM): biological sequence clustering and cluster visualization. Methods Mol. Biol. **1552**, 83–101 (2017)
2. Cole, J.R., et al.: Ribosomal Database Project: data and tools for high throughput rRNA analysis. Nucleic Acids Res. **42**(Database issue), D633–D642 (2014). https://doi.org/10.1093/nar/gkt1244
3. Lu, X.X., Wu, W., Wang, M., Huang, Y.F.: 16S rRNA gene sequencing for pathogen identification from clinical specimens. Zhonghua Yi Xue Za Zhi **88**(2), 123–126 (2008). https://doi.org/10.3321/j.issn:0376-2491.2008.02.014
4. Gales, M., Young, S.: The application of hidden Markov models in speech recognition. Found. Trends Sig. Process. **1**(3), 195–304 (2008). https://doi.org/10.1561/2000000004

5. Yoon, B.-J.: Hidden Markov models and their applications in biological sequence analysis. Curr. Genomics **10**(6), 402–415 (2009). https://doi.org/10.2174/138920209789177575

6. Xing, Z., Jian, P., Eamonn, K.: A brief survey on sequence classification. SIGKDD Explor. **12**(1), 40–48 (2010). https://doi.org/10.1145/1882471.1882478

7. Kang, M.-S., Kim, H., Lee, S., Kim, M.H.: Feature-based gene classification and region clustering using gene expression grid data in mouse Hippocampal region. J. KIISE **43**(1), 54–60 (2016). https://doi.org/10.5626/JOK.2016.43.1.54

8. Hawrylycz, M., et al.: Multi-scale correlation structure of gene expression in the brain. Neural Netw. **24**(9), 933–942 (2011)

9. Chandra, B., Gupta, M.: An efficient statistical feature selection approach for classification of gene expression data. **44**(4), 529–535 (2011). https://doi.org/10.1016/j.jbi.2011.01.001

10. Abusamra, H.: A comparative study of feature selection and classification methods for gene expression data of glioma, 5–14 (2013). https://doi.org/10.1016/j.procs.2013.10.003

11. Doungpan, N., Engchuan, W., Meechai, A., Fong, S., Chan, J.H.: Gene-Network-Based Feature Set (GNFS) for expression-based cancer classification. J. Med. Imaging Health Inform. **6**(4), 1093–1101 (2016). https://doi.org/10.1166/jmihi.2016.1806

12. Baralis, E., Bruno, G., Fiori, A.: Measuring gene similarity by means of the classification distance. Knowl. Inf. Syst. **29**(1), 81–101 (2011)

13. Iqbal, M.J., Faye, I., Said, A.M., Belhaouari Samir, B.: A distance-based feature-encoding technique for protein sequence classification in bioinformatics. In: IEEE International Conference on Computational Intelligence and Cybernetics (CYBERNETICSCOM), pp. 1–5 (2013). https://doi.org/10.1109/CyberneticsCom.2013.6865770

14. Kaya, H., Gunduz Oguducu, S.: A distance based time series classification framework. Inf. Syst. (2015). https://doi.org/10.1016/j.is.2015.02.005

15. Chen, H., Zhang, Y., Gutmanb, I.: A kernel-based clustering method for gene selection with gene expression data. J. Biomed. Inform. 12–20 (2016). https://doi.org/10.1016/j.jbi.2016.05.007

16. Wang, S., Li, X., Zhang, S.: Neighborhood rough set model based gene selection for multi-subtype tumor classification. In: Huang, D.-S., Wunsch, D.C., Levine, D.S., Jo, K.-H. (eds.) ICIC 2008. LNCS, vol. 5226, pp. 146–158. Springer, Heidelberg (2008). https://doi.org/10.1007/978-3-540-87442-3_20

17. Bauer, S., Robinson, P.N., Gagneur, J.: Model-based gene set analysis for Bioconductor. Bioinformatics **27**(13), 1882–1883 (2011). https://doi.org/10.1093/bioinformatics/btr296

18. Bauer, S., Gagneur, J., Robinson, P.N.: Going Bayesian: model-based gene set analysis of genome-scale data. Nucleic Acids Res. **38**(11), 3523–3532 (2010). https://doi.org/10.1093/nar/gkq045

19. Guo, P., et al.: Gene expression profile based classification models of psoriasis. Genomics **103**(1), 48–55 (2014). https://doi.org/10.1016/j.ygeno.2013.11.001

20. Onan, A., Korukolu, S.: A feature selection model based on genetic rank aggregation for text sentiment classification. **43**(1), 25–38 (2015). https://doi.org/10.1177/0165551515613226

21. Saengsiri, P., Meesad, P., Wichian, S.N., Herwig, U.: Classification models based-on incremental learning algorithm and feature selection on gene expression data. In: 8th Electrical Engineering/Electronics, Computer, Telecommunications and Information Technology (ECTI) Association of Thailand - Conference, pp. 426–429 (2011). https://doi.org/10.1109/ECTICON.2011.5947866

22. Welch, L.: Hidden Markov models and the Baum-Welch algorithm. IEEE Inf. Theory Soc. Newsl. **53**(4), 10–13 (2003)

23. Karplus, K., et al.: Predicting protein structure using hidden Markov models. Proteins **1**, 134–139 (2007)

24. Yakhnenko, O., Silvescu, A., Honavar, V.: Discriminatively trained Markov model for sequence classification. In: Fifth IEEE International Conference on Data Mining, pp. 1–8 (2005). https://doi.org/10.1109/ICDM.2005.52

25. Srivastava, P.K., Desai, D.K., Nandi, S., Lynn, A.M.: HMM-ModE-Improved classification using profile hidden Markov models by optimizing the discrimination threshold and modifying emission probabilities with negative training sequences. BMC Bioinform. (2007). https://doi.org/10.1186/1471-2105-8-104

26. Camproux, A.C., Tuffery, P., Chevrolat, J.P., Boisvieux, J.F., Hazout, S.: Hidden Markov model approach for identifying the modular framework of the protein backbone. Protein Eng. **12**(12), 1063–1073 (1999)

27. Sonnhammer, E.L.L., Eddy, S.R., Birney, E., Durbin, R.: Multiple sequence alignments and HMM-profiles of protein domains. Nucleic Acids Res. **26**(1), 320–322 (1998)

28. Di Francesco, V., Garnier, J., Munson, P.J.: Protein topology recognition from secondary structure sequences: application of the hidden Markov models to the alpha class proteins. J. Mol. Biol. **267**(2), 446–463 (1997)

29. Liu, T., Lemeire, J., Yang, L.: Proper initialization of Hidden Markov models for industrial applications. In: IEEE China Summit and International Conference on Signal and Information Processing (ChinaSIP), pp. 490–494 (2014). https://doi.org/10.1109/ChinaSIP.2014.6889291

30. Mann, T.P.: Numerically stable Hidden Markov Model implementation (2006)

31. Tatavarty, U.R.: Implementation of numerically stable hidden Markov model. UNLV Theses, Dissertations, Professional Papers, and Capstones. 1018 (2011). http://digitalscholarship.unlv.edu/thesesdissertations/1018

32. Fu, B.: Computer architecture. Fall Project Report (2009)

33. Jose, S., Nair, P., Biju, V.G., Mathew, B.B., Prashanth, C.M.: Hidden Markov model: application towards genomic analysis. In: International Conference on Circuit, Power and Computing Technologies (ICCPCT), pp. 1–7. IEEE (2016). https://doi.org/10.1109/ICCPCT.2016.7530222

34. Vijayabaskar, M.S.: Introduction to hidden Markov models and its applications in biology. In: Westhead, D.R., Vijayabaskar, M.S. (eds.) Hidden Markov Models: Methods and Protocols, Methods in Molecular Biology, vol. 1552 (2017)

35. Wang, Q., Garrity, G.M., Tiedje, J.M., Cole, J.R.: Nave Bayesian classifier for rapid assignment of rRNA sequences into the new bacterial taxonomy. Appl. Environ. Microbiol. **73**(16), 61–67 (2007)

36. Ghosh, T.S., Gajjalla, P., Mohammed, M.H., Mande, S.S.: C16S A Hidden Markov Model based algorithm for taxonomic classification of 16S rRNA gene sequences. Genomics **99**(4), 195–201 (2012). https://doi.org/10.1016/j.ygeno.2012.01.008

37. Janda, J.M., Abbott, S.L.: 16S rRNA gene sequencing for bacterial identification in the diagnostic laboratory: Pluses, Perils, and Pitfalls. J. Clin. Microbiol. **45**(9), 2761–2764 (2007). https://doi.org/10.1128/JCM.01228-07

38. Fontana, C., Favaro, M., Pelliccioni, M., Pistoia, E.S., Favalli, C.: Use of the MicroSeq 16S rRNA gene based sequencing for identification of bacterial isolates that commercial automated systems failed to identify correctly. J. Clin. Microbiol. **43**(2), 615–619 (2005)
39. Patel, J.B.: 16S rRNA gene sequencing for bacterial pathogen identification in the clinical laboratory. Mol. Diagn. **6**(4), 313–321 (2001)
40. Mizrahi-Man, O., Davenport, E.R., Gilad, Y.: Taxonomic classification of bacterial 16S rRNA genes using short sequencing reads: evaluation of effective study designs. PLoS ONE **8**(1), e53608 (2013). https://doi.org/10.1371/journal.pone.0053608
41. Song, Y., Liu, C., BolaÅos, M., Lee, J., McTeague, M., Finegold, S.M.: Evaluation of 16S rRNA sequencing and reevaluation of a short biochemical scheme for identification of clinically significant Bacteroides species. J. Clin. Microbiol. **43**(4), 1531–1537 (2005)
42. Heikens, E., Fleer, A., Paauw, A., Florijn, A., Fluitt, A.C.: Comparison of genotypic and phenotypic methods for species-level identification of clinical isolates of coagulase-negative staphylococci. J. Clin. Microbiol. **43**(5), 2286–2290 (2005)
43. Bosshard, P.P., Zbinden, R., Abels, S., Bddinghaus, B., Altwegg, M., Bttger, E.C.: 16S rRNA gene sequencing versus the API 20 NE system and the VITEK 2 ID-GNB card for identification of nonfermenting Gram-negative bacteria in the clinical laboratory. J. Clin. Microbiol. **44**(4), 1359–1366 (2006)

BER$_y$L: A System for Web Block Classification

Andrey Kravchenko$^{(\boxtimes)}$

Department of Computer Science, University of Oxford, Oxford, UK
andrey.kravchenko@cs.ox.ac.uk

Abstract. Web blocks such as navigation menus, advertisements, headers, and footers are key components of Web pages that define not only the appearance, but also the way humans interact with different parts of the page. For machines, however, classifying and interacting with these blocks is a surprisingly hard task. Yet, Web block classification has varied applications in the fields of wrapper induction, assistance to visually impaired people, Web adaptation, Web page topic clustering, and Web search. Our system for Web block classification, BER$_y$L, performs automated classification of Web blocks through a combination of machine learning and declarative, model-driven feature extraction based on Datalog rules. BER$_y$L uses refined feature sets for the classification of individual blocks to achieve accurate classification for all the block types we have observed so far. The high accuracy is achieved through these carefully selected features, some even tuned to the specific block type. At the same time, BER$_y$L avoids a high cost of feature engineering through a model-driven rather than programmatic approach to extracting features. Not only does this reduce the time for feature engineering, the model-driven, declarative approach also allows for semi-automatic optimisation of the feature extraction system. We perform evaluation to validate these claims on a selected range of Web blocks.

1 Introduction

When a human looks at a Web page, he or she sees a meaningful and well-structured document. However, such interpretation is not accessible for the computer as it only "sees" the technical layers of the Web page [19] represented merely by the source code (e.g. HTML, CSS, and JavaScript files) and the rendered models (e.g. the DOM tree, CSSOM with computed attributes, and executed JavaScript). Whilst it is probably infeasible for a machine to replicate the human's perception and derive the associated mental model, it would be highly useful for it to understand the logical structure and functional role of various elements of the Web page for a wide range of different applications through

This work was supported by the ESPRC programme grant EP/M025268/1 "VADA: Value Added Data Systems – Principles and Architecture".

© Springer-Verlag GmbH Germany, part of Springer Nature 2018
M. L. Gavrilova and C. J. K. Tan (Eds.): Trans. on Comput. Sci. XXXIII,
LNCS 10990, pp. 61–78, 2018.
https://doi.org/10.1007/978-3-662-58039-4_4

the analysis of the layout, as well as visual and textual features. Web search is an especially important potential application, since semantic understanding of a Web page allows the restriction of link analysis to clusters of semantically coherent blocks. Hence, we aim to build a system which provides a structural and semantic understanding of Web pages.

This paper is concerned with the task of Web block classification. Informally speaking, a Web block is a logically consistent segment of a Web page layout, an area which can be identified as being visually separated from other parts of the Web page. A Web block carries a certain semantic meaning, such as title, main content, advertisement, login area, footer, and so on. It is through the semantic meaning of individual Web blocks that a human understands the overall meaning of a Web page. There are many blocks with a common semantic meaning (i.e. a layer of Web specific objects [19]) among different websites and domains (e.g. headers, navigation menus, logos, pagination elements, and maps) that share common Web patterns. Even within one block type, individual blocks can vary significantly in both their structural and visual representations. For example, consider the diversity of navigation menus illustrated in Fig. 1. This diversity of blocks makes the task of their accurate and fast detection challenging from a research and implementation perspective. In general, the difficulty of the block classification problem lies not only in the complexity of individual classifiers, but also in the complexity of the entire system which needs to balance the individual accuracies of its constituent classifiers and its overall performance.

There are several important applications of Web block classification including automatic and semi-automatic wrapper induction [2,11,28,35], assisting visually impaired people with navigating the website's internal content [14], help in the task of mobile Web browsing [3,14,24,25,30,31], ad removal, Web page topic clustering [23], and the ubiquitous task of a Web search [5,6,27,29,34].

Our Web block classification system is known as BER$_y$L (**B**lock classification with **E**xtraction **R**ules and machine **L**earning). There are three main requirements that it must meet:

1. be able to cover a diverse range of blocks;
2. achieve acceptable precision and recall results for each individual block in the classification system, and maximise the overall performance of the system;
3. be adaptive to new block types and domains.

The task of Web block classification is technically challenging due to the diversity of blocks in terms of their internal structure (their representation in the DOM tree and the visual layout) and the split between domain-dependent and domain-independent blocks. Hence, from a technical perspective, it is important to have a global feature repository which can provide a framework for defining new features and block type classifiers through the instantiation of template relations. A BER$_y$L user will be able to extend it with new features and classifiers with ease by generating them from existing template relations, rather than writing them from scratch. This will make the whole hierarchy of features and classifiers leaner and the process of defining new block types and their respective classifiers more straightforward and less time-consuming. Ideally, we aim to generate new block classifiers in a fully automated way such that, given a set of structurally and visually distinct Web blocks of the same type, the block classification system would be able to automatically identify the list of optimal features for describing that block, taking some of these features from the existing repository and generating new ones which did not exist in the repository beforehand. However, this approach would almost be infeasible in the case of BER$_y$L since the diversity of block types that we want to classify is likely to cause the space of potential features to be extraordinarily large, if not infinite. Hence, we will need to limit the approach of the generation of new features and block type classifiers to a semi-automated approach.

Fig. 1. The diversity of navigation menu types

The contributions of our approach in BER$_y$L have two main aims: improving the quality of classification and minimising the effort of generating new classifiers (as explained in the above paragraph). Contributions 1–2 and 3–4 refer to the quality of classification and generation aspects respectively.

1. We employ domain-specific knowledge to enhance the accuracy and performance of Web block classifiers.
2. We provide template and global feature repositories which allow users of BERyL to easily add new features and block classifiers.
3. We encode domain-specific knowledge through a set of logical rules, e.g. that the navigation menu is always at the top or bottom of (and rarely to the side of) the main content area of the page.

4. The template and global feature repositories are implemented through baseline global features and template relations used to derive local block-specific features.

With respect to the employment of domain-specific knowledge for enhancing the performance of classifiers (contributions 1 and 3), our new approach to feature extraction allows us to easily integrate domain-specific pre- and post-classification filters which will ensure that the classifiers in question meet all the additional restrictions imposed by the domain in which they are applied.

We have also implemented template and global feature repositories (contributions 2 and 4) which are crucial to the automation and large-scale evaluation of the BER_yL system. The template repository is implemented through component-driven approach to feature extraction that is covered in detail in Sect. 3. Our system also supports baseline global features which can be shared between different classifiers. The template repository and baseline global features allow us to easily extend the system with new classifiers without reducing the accuracy of the classifiers that are already present in the system or significantly reducing the system's performance.

2   Related Work

There have been several papers published on this topic in the past ten years including [4,7,8,12,15–17,20–22,24,26,32,33,35,36]. Broadly speaking, Web block classification methods can be split into those based on machine learning (ML) techniques and those based on other approaches (i.e. the rule-based and heuristics). Also, the features used for the analysis of Web blocks in ML-driven and other algorithms can be subdivided into structural (based on the structure of a DOM tree or another intermediary structure representing a Web page), visual (based on the Web page's rendering), and lexical (based on the analysis of the page's free text).

Most of the approaches taken in this research have attempted to classify a relatively small set of domain-independent blocks with a limited number of features, whereas we aim to develop a unified approach to classify both domain-dependent and domain-independent blocks. Furthermore, none of the papers with which we are familiar discussed the extensibility of their approaches to new block types and features.

Finally, although most of these machine learning-based approaches require significantly more training than BER_yL, they usually reach a precision level below 70% [12,15,17,20,24]. The highest value of F_1 that any of these machine learning-based approaches reach (apart from [10] that reaches the F_1 of 99% but for a single very specific block type of "next" links and [18] that reaches the overall classification rate of 94% but for basic Web form elements within the transportation domain and requires a labour-intensive labelling and complex training procedure) is around 85% [24]. An evaluation of our BER_yL system (Sect. 4) has shown that it can achieve much higher levels of precision and recall. This can partly be explained by the fact that they have attempted to classify different blocks with the same set of features, whereas we attempt to employ individual feature sets for different types of blocks.

Although most of the current block classification approaches are based on machine learning methods, to our knowledge there is no universal approach to the task of block classification.

3 Approach

The problem of Web block classification is, given a Web page P with associated annotated DOM T_P, to find a mapping from the set of sub-trees $T'_1, \ldots, T'_n$ of T_P to sets of labels, such that each sub-tree is labelled with, at most, one label per block type. The novelty of our approach is that we try to tailor feature sets for different Web blocks by application of declarative programming to feature extraction. We propose a component-driven approach combined with the use of Datalog-based BER$_y$L language and powered by the use of block-specific knowledge. This, together with the repository of global features and relation templates, helps to create tailored features for different blocks and domains.

Component-Based Approach to Feature Extraction. We give a description of how our approach to feature extraction works in practice by introducing the concept of a **Component Model (CM)-driven approach to feature extraction**. Components are fragments of code that can be combined to allow for modular definition of complex features that we want to extract. Components can have different types (e.g. we distinguish between rule-based components that are mapped to Datalog programs and procedural components that are mapped to fragments of Java code). Components can also have parameters attached to them, which can either be atomic parameters, e.g. font size, sequential parameters, e.g. the tags of the DOM tree we want to consider for analysis, or higher-level parameters that can link to other components within the definition of a given component. Parameters are assigned values through *instantiations* $\mathfrak{C}[p \mapsto v]$, and for higher-level parameters we allow *references* as values, i.e. expressions of the shape @n where n is the unique name of the referenced component.

Definition 1 (Component type). A *component type* $\mathfrak{C}$ is a triple $\langle I, O, U \rangle$ of two relation schemas I and O, the respective schema of the input and output relations for a component, and a set of formal parameters U.

Definition 2. A *formal component parameter* is a mapping of a unique name p, the parameter name, to a parameter type $\tau_p \in \mathscr{P}$. $\mathscr{P}$ is the set of permissible types and defined recursively as follows: Let $\mathscr{P}_A$ be the set of *atomic* types such as *Integer* or *String*. Then a permissible type is either an atomic type from $\mathscr{P}_A$, a sequence type $[\tau_i]$ where τ_i is itself another type from $\mathscr{P}$, or a component type.

$$\tau_p = \begin{cases} \alpha & \text{with } \alpha \in \mathscr{P}_A \\ [\tau_i] & \text{with } \tau_i \in \mathscr{P} \\ \mathfrak{C} & \text{with } \mathfrak{C} \text{ a component type} \end{cases}$$

For each type τ, we denote with $\mathsf{domain}(\tau)$ the set of permissible values for τ.

We call a parameter of atomic type an atomic parameter, of sequence type a sequence parameter, and of component type a component parameter. For an atomic type τ, $\mathsf{domain}(\tau)$ is the corresponding set of values. For a sequence type, it is the set of sequences over the values of the constituent type. For a component type, it is the set of (references to) ground components of type $\mathfrak{C}$. We call a component *generic* if it has at least one formal parameter, and *higher-order* if there is at least one component type parameter in U.

Let $\mathcal{N}$ be the set of component names. Then $\mathfrak{C}[p \mapsto v]$ is an *instantiation* for component type $\mathfrak{C}$ that binds the formal parameter with name p in $\mathfrak{C}$ to the value v, where $v \in \mathsf{domain}(p)$. We call an instantiation ground if all parameters in $\mathfrak{C}$ are bound to an actual value. For component parameters, we allow *references* as values, i.e. expressions of the shape @n where n is the unique name of the referenced component.

Definition 3. Let $\mathcal{K}$ be the set of primitive components. A *(component) composition specification* describes how to combine components into more complex ones. A composition specification for a component type $\mathfrak{C}$ is then a triple $C = (n, \mathfrak{C}', \mathsf{E})$ where n is a unique name, $\mathfrak{C}'$ a ground instantiation for component type $\mathfrak{C} = (I, O, U)$, and E an expression that describes how to compose this component from other, more primitive ones. Specifically, a composition expression is an expression of the form:

$$\mathsf{E} = \begin{cases} k & \text{where } k \in \mathcal{K} \\ @n & \text{where } n \text{ is a reference to an already} \\ & \text{defined component} \\ \mathsf{E}_1 \lhd \ldots \lhd \mathsf{E}_r & \text{where } \mathsf{E}_i \text{ are composition expressions} \\ \mathsf{E}_1 \parallel \ldots \parallel \mathsf{E}_r & \text{where } \mathsf{E}_i \text{ are composition expressions} \\ \mathsf{E}_1 \oplus \ldots \oplus \mathsf{E}_r & \text{where } \mathsf{E}_i \text{ are composition expressions} \end{cases}$$

We call $\lhd$ isolated *sequential* composition, $\parallel$ *parallel*, but isolated composition, and $\oplus$ sequential, but *possibly dependent* composition.

We write $\mathrm{TYPE}(C) = \mathfrak{C}$, $\mathrm{INPUT}(C) = I$, and $\mathrm{OUTPUT}(C) = O$. Component expressions form trees (with possibly shared branches) and thus no cycles between components are possible. We refer to the name n assigned to a composition expression e as $\mathrm{NAME}(e)$.

We refer to the components that are referenced in a composition expression as sub-components $C_1, \ldots, C_r$ and r the arity of C. We call the set of *defined sub-components* in E_C the set $\mathrm{SUBS}(C)$ of all components defined in E_C or one of its sub-expressions, i.e. $\mathrm{SUBS}(C) = \bigcup_{1 \leq i \leq r} \mathrm{SUBS}(C_i) \cup \{(n_i : C_i) : 1 \leq i \leq r\}$.

A composition expression E_C is *valid*, if

1. the schemata of the sub-components are compatible: $\mathrm{INPUT}(C) = \mathrm{INPUT}(n_1)$, $\mathrm{OUTPUT}(n_{i-1}) = \mathrm{INPUT}(n_i)$ for all $1 < i \leq r$, and $\mathrm{OUTPUT}(n_r) = \mathrm{OUTPUT}(C)$.

2. for each composition parameter $(p, \tau) \in \text{TYPE}(C)$ that is instantiated to component reference @v, there is a sub-component $v : C_v \in \text{SUBS}(C)$ such that $\text{TYPE}(C) = \tau$.

Given a component (and corresponding composition expression) the semantic of that component is then quite straightforward: in our framework, components are implemented either as declarative rules (rule-based components, written and evaluated as Datalog rules) or Java classes (procedural components). In both cases, implementations may query parameters of the surrounding component. For atomic parameters, the query returns the corresponding value, for sequence parameters it returns a suitable representation of a sequence in Datalog or Java, and for component parameters it returns an interface that allows the implementation to access the results of the other component.

Definition 4. Let $C = (n, \mathfrak{C}')$ be a component with corresponding composition expression $n : \mathsf{E}_C$ and U the set of parameters for $\text{TYPE}(C)$. Then, we denote with $[\![\mathsf{E}_C]\!](I)$ the output for C under E if executed on input I. The output is a set of pairs $n : O$ where n is a component name and O an instance of a component output schema. As such, we write $[\![\mathsf{E}_C]\!](I)[n]$ to access the output instance for the component with name n.

$$[\![n : \mathsf{E}_C]\!](I) = \{n : [\![\mathsf{E}_C]\!](I \cup U)[\text{NAME}(\mathsf{E}_C)]\}$$

$$[\![\mathsf{E}_C(I)]\!] = \begin{cases} \{n_1 : k(I)\} & \text{if } \mathsf{E} = k \in \mathscr{K} \\ \bigcup_{i \leq r}[\![C_i]\!](I) & \text{if } \mathsf{E} = C_1 \parallel \ldots \parallel C_r \\ \bigcup_{i \leq r}[\![C_r]\!]([\![C_{r-1}]\!](\ldots [\![C_1]\!](I)\ldots)[\text{NAME}(C_{r-1})]) & \text{if } \mathsf{E} = C_1 \lhd \ldots \lhd C_r \\ \bigcup_{i \leq r}[\![C_r]\!]([\![C_{r-1}]\!](\ldots [\![C_1]\!](I)\ldots) \cup I) & \text{if } \mathsf{E} = C_1 \oplus \ldots \oplus C_r \end{cases}$$

We can now give a definition of a *component model*, which is a crucial part of our BER$_y$L system, since each of its constituent classifiers corresponds to a unique component model.

Definition 5 (Component model). A component model is a set M of executable components such that for each component $C \in \mathsf{M}$ with a higher-order parameter p that references a component C', then $C' \in \mathsf{M}$ ("no dangling references").

Example 1. We give an example of a fragment of the real component model we use for the Next Link classifier (Fig. 2). Let us consider one of the most important features of the Next Link classifier that checks whether the direct left numeric visual sibling of a given numeric node is a link or not (if it is not a link that is a strong indicator that the numeric link in question is a numeric next link). We define this feature from the global template relation *two nodes with unary relation properties connected by a binary relation* that denotes two nodes connected to each other by a binary relation that also defines a unary property of the second node. Each of the nodes also holds one unary relation property of their own. That is a very generic template that can be used in

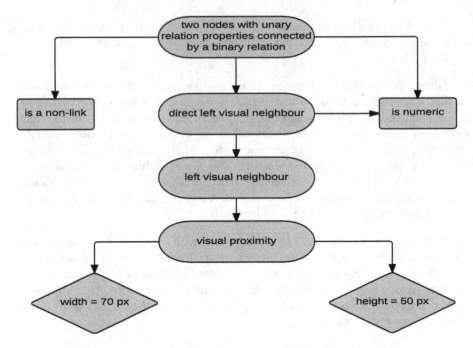

Fig. 2. An example of a component model for a single feature

many cases for different classifiers. In our case, the binary relation is the *direct left visual neighbour* relation with the property of the second node (the one to the visual left of the node in question) defined as *is numeric*. Note that the *direct left visual neighbour* binary relation, in turn, utilises the *visual proximity* binary relation that defines any visual proximity within the given boundaries regardless of direction. We set the width and height parameters for the visual proximity boundaries to 70 and 50 pixels respectively. The respective unary relation properties of the two nodes in question are *is numeric* for the first node and *is a non-link* for the second node.

The BER$_y$L *Language.* We provide a tool to the users of our system that allows them to define powerful BER$_y$L component models in an easy and intuitively understandable, declarative way. This BER$_y$L language is a version of Datalog with some extensions and is one of the key modules of our framework.

The BER$_y$L language is a dialect of Datalog that includes safe and stratified negation, aggregation, arithmetic, and comparison operators, denoted as Datalog$^{\neg,Agg,ALU}$. The usual restrictions apply, i.e. all variables in a negation, arithmetic operation, aggregation, or comparison must also appear in some other (positive) context. Further, there may be no dependency cycles over such operators. In the context of the BER$_y$L system, this language is used in a certain way:

Table 1. BER$_y$L's input facts

Structural:

`dom::element(N,T,PStart,Start,End)`	DOM element node N $\in \mathcal{N}$ has tag T=τ(N), and parent node P, such that (P, N) $\in$ CHILD, with start label PStart, and start and end labels Start and End.[1]
`dom::attribute(A,N,T,V)`	DOM attribute node A of node N with (N, A) $\in$ ATTR has tag T=τ(A) and value V=σ(A).
`dom::clickable(N)`	N is a clickable target (a link or has an onclick handler, i.e. (τ(N)=a) $\vee$ ($\exists$(N,A) $\in$ ATTR $\wedge$ τ(A)=onclick).
`dom::content(N,v,O,L)`	Node N corresponds to textual content v with (N, N$_i$) $\in$ CHILD$^+$ $\wedge \neg \exists$ (N$_i$, N$_j$) $\in$ CHILD $\wedge$ (σ(N$_1$)++...++σ(N$_n$)=v)[2], nodes N$_i$ are sorted in the ascending order of their Start labels, node N starts at document offset O and has length L.

Visual:

`css::box(N,Left,Top,Right,Bottom)`	bounding box of a DOM node N, such that $\exists$(N,C$_1$), (N,C$_2$), (N,C$_3$), (N,C$_4$) $\in$ CSS: τ(C$_1$)=left_coord$\wedge\tau$(C$_2$)=top_coord$\wedge$ τ(C$_3$)=right_coord$\wedge\tau$(C$_4$)=bottom_coord$\wedge$ σ(C$_1$)=Left$\wedge\sigma$(C$_2$)=Top$\wedge$ σ(C$_3$)=Right$\wedge\sigma$(C$_4$)=Bottom.
`css::page(Left,Top,Right,Bottom)`	bounding box of a DOM node N with τ(N)=html.
`css::resolution(1800,800)`	the average screen resolution.[3]
`css::font_family(N,Family)`	N is rendered with a font from the given family ($\exists$(N,C) $\in$ CSS $\wedge \tau$(C)=font_family$\wedge\sigma$(C)=Family).
`css::font_size(N,Size)`	N is rendered with a font of the given size ($\exists$(N,C) $\in$ CSS $\wedge \tau$(C)=font_size$\wedge\sigma$(C)=Size).

Content:

`gate::annotation(N,A,v)`	the text of node *N* contains an instance of entity type *A* and the normalised representation of that instance is *v*.

Classification:

`cls::classification(N,BT,Label)`	a classification (N, Label) $\in$ CLASS where *BT* $\in$ $\mathbb{BT}$ is a block type with Label$\in$ *BT*. For instance, the pagination classifier corresponds to the `plm` (pagination link model) namespace, e.g. `cls::classification(e_200,pagination, numeric_next_link)` and `cls::classification(e_300,pagination,non_num_next_link)`. Different classifiers can produce output on the same node, e.g. `cls::classification(e_100,header,header)` and `cls::classification(e_100,navigation, nav_menu)`.

---

[a]Start and end labels of a node correspond to a pre-order traversal of the DOM tree with a single incremental counter that assigns the start label the first time the node has been explored, and the end label when all the node's descendants have been explored.
[b]++ is a concatenation operator.
[c]According to a study by http://www.w3schools.com/browsers/browsers_display.asp

1. there is a number of input facts for representing the annotated DOM (see Table 1), as well as the output of previous components, such as the classification facts in Table 1;
2. each program carries a distinguished set of output predicates and only entailed facts for those predicates are ever returned.

The BER_yL language also uses a number of syntactic extensions to ease the development of complex rule programs. The most important ones are *namespaces*, *parameters*, and the explicit distinction of *output predicates*. All other intensional predicates are temporary ("helper") predicates only. For the purpose of brevity, we omit the precise definitions of these extensions.

Specifically, there are two classes of components and component types, which are used heavily in BER_yL language programs: components representing (1) relations and (2) features. These components operate on the universe $\mathscr{U}$ of DOM nodes and atomic values appearing in the DOM of a page. We typically use relations to either (a) distinguish sets of nodes, (b) relate sets of nodes to each other, or (c) attach additional information to nodes.

The Standard Library of the BER_yL *Language.* The BER_yL language also provides a set of predefined components that a specific block type may import, which are defined through the input facts from Table 1 and are written in the `relation` namespace. We call these predefined components **standard relations**. Figure 3 shows some of the standard relations in BER_yL: these range from structural relations between nodes (similar to XPath relations) over visual relations (such as proximity) to information about the rendering context (such as the dimensions of the first and last screen).

Example 2. Let us reconsider Example 1. The graph representation of the component model for this feature is given in Fig. 2. For the purpose of conciseness in our notation we abbreviate the main component and its constituent subcomponents in the following way: (a) is numeric as C_1, (b) is a non-link (or non-clickable) as C_2, (c) visual proximity as C_3, (d) left visual neighbour as C_4, (e) direct left visual neighbour as C_5, and (f) two nodes with unary relation properties connected by a binary relation as C_6. C_1–C_2 are unary relation components, whilst C_3–C_6 are binary relation components. We give the rules for these components in Fig. 4.

C_1 and C_2 are primitive components with no parameters attached to them and therefore have trivial empty instantiations and identity-binding expressions:

```
  𝕮₁ [] ;  C₁
 2 𝕮₂ [] ;  C₂
```

The rules of C_3 and C_4 correspond to standard binary relations `relation::visual_proximity` and `relation::left_visual_neighbour` as defined in Fig. 3. C_3 has two parameters dH and dV that represent the possible positions of the top-left coordinate of CSS boxes corresponding to node considered to be in the visual proximity of the CSS box of the given node. The instantiations and composition expressions of these two components are the following:

```
  relation::preceding(Id,X,Y) ⇐
2   dom::element(X,_,_,_,End),
    dom::element(Y,_,_,Start,_), End < Start,
4   param::instantiation_id(Id).
  relation::descendant(Id,X,Y) ⇐
6   dom::element(X,_,_,StartX,EndX),
    dom::element(Y,_,_,StartY,EndY), StartY < StartX,
8   EndX < EndY,
    param::instantiation_id(Id).
10 relation::leaf_descendant(Id,X,Y) ⇐ dom::element(X,_,_,_,_),
    relation::descendant(CId₁,X,Y),
12  ¬relation::descendant(CId₂,Z,X),
    param::instantiation_id(Id).
14 relation::visual_proximity(Id,X,Y) ⇐
    css::box(X,LeftX,TopX,_,_),
16  css::box(Y,LeftY,TopY,_,_),
    TopY-DVert ≤ TopX ≤ TopY+DVert,
18  LeftY-DHor ≤ LeftX ≤ LeftY+DHor,
    param::dH(PId₁,DHor), param::dV(PId₂,DVert),
20  param::instantiation_id(Id).
  relation::left_visual_neighbour(Id,X,Y) ⇐
22  relation::visual_proximity(CId,X,Y),
    css::box(X,_,_,RightX,_), css::box(Y,LeftY,_,_,_),
24  RightX ≤ LeftY, param::instantiation_id(Id).
  relation::first_screen(Id,Left,Top,Right,Bottom) ⇐
26  css::page(Left,Top,_,_), css::resolution(H,V),
    Right = Left+H, Bottom = Top+V, param::instantiation_id(Id).
28 relation::last_screen(Id,Left,Top,Right,Bottom) ⇐
    css::page(_,_,Right,Bottom), css::resolution(H,V),
30  Left = Right-H, Top = Bottom-V, param::instantiation_id(Id).
```

Fig. 3. The standard library of the BER$_y$L language

$\mathfrak{C}_3$[dH↦70,dV↦50]; C_3
2 $\mathfrak{C}_4$[]; $C_3 \triangleleft C_4$

We now give the instantiation and the binding expression of binary relation component C_5 that specifies whether one node is a direct left visual neighbour of another node and has a single unary relation component parameter:

$\mathfrak{C}_5$[sibling_pred↦ C_1]; $((C_3 \triangleleft C_4) \parallel C_1) \triangleleft C_5$

Finally, we give the instantiation and binding expression of binary relation component C_6 according to our example:

$\mathfrak{C}_6$[binary_pred↦ C_5,node_pred↦ C_1,sibling_pred↦ C_2];
2 $((((C_2 \parallel C_3) \oplus C_4) \oplus C_5) \parallel C_2) \triangleleft C_6$

Note that it seems more obvious to define the composition expression for C_5 as $((C_1 \parallel C_3) \triangleleft C_4) \triangleleft C_5$, but in that case we would have to recompute

```
  C₁: relation::numeric(Id₁,X) ⇐ gate::annotation(X,"NUMBER",_),
2      param::instantiation_id(Id₁).
  C₂: relation::non_clickable(Id₂,X) ⇐ dom::element(X,_,_,_,_),
4      ¬(dom::clickable(X)), param::instantiation_id(Id₂).
  C₃: relation::visual_ proximity(Id₃,X,Y)
6 C₄: relation::left_visual_neighbour(Id₄,X,Y)
  C₅: relation::direct_left_visual_neighbour(Id₅,X,Y) ⇐
8      relation::left_visual_neighbour(CId₁,X,Y),
       ¬relation::indirect_left_visual_neighbour(CId₂,X,Y),
10     param::sibling_pred(CId₃,Y),
       param::instantiation_id(Id₅).
12     relation::indirect_left_visual_neighbour(CId₂,X,Y) ⇐
       relation::left_visual_neighbour(CId₄,X,Y),
14     relation::left_visual_neighbour(CId₅,X,Z),
       css::box(Y,LeftY,_,RightY,_), css::box(Z,LeftZ,_,RightZ,_),
16     RightY ≤ LeftZ, RightX ≤ LeftY,
       param::instantiation_id(CId₂).
18 C₆: relation::binary_unary(Id₆,X,Y) ⇐
       param::binary_pred(CId₁,X,Y),
20     param::node_pred(CId₂,X),
       param::sibling_pred(CId₃,Y),
22     param::instantiation_id(Id₆).
```

Fig. 4. Rules corresponding to components C_1–C_6

C_1 at the time of evaluation of the final component C_6, and the composition expression for C_6 would have been $((((C_1 \parallel C_3) \lhd C_4) \lhd C_5) \parallel (C_1 \parallel C_2)) \lhd C_6$, as the semantics of the $\lhd$ operator would have restricted visibility of C_2 at the time of the computation of C_6, and we would have encountered an additional computational overhead.

Declarative Approach to Feature Extraction. In BER$_y$L, we use a declarative approach to feature extraction wherever possible, in particular through the BER$_y$L language described above, since that (**1**) allows us to combine it with relations and global features which provide a succinct representation of the current feature set. This, in turn, allows us to simplify the definition of new features through the employment of existing global features or relation instantiations and to learn new features by automatically finding the right combination of parameters for relation instantiations. (**2**) It is also much easier to learn Datalog [1,9] predicates automatically than to learn procedural language programs, which is likely to come in useful in the large-scale block classification phase of BER$_y$L when we will have to infer new features automatically. In other cases which require the use of efficient libraries and data structures or intense numerical computation (e.g. features acquired from image processing), we employ a procedural approach for feature extraction implemented through Java.

We use Datalog as the declarative language of choice, since it is fast and widely used [9,13]. Also, with a Datalog-based approach to feature extraction we

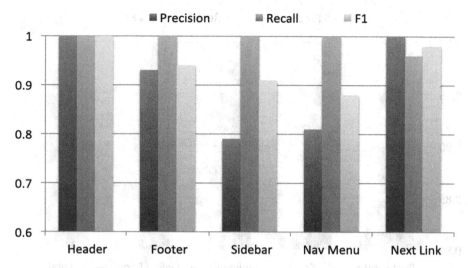

Fig. 5. BER$_y$L's accuracy results on the five classifiers

can easily extend our extraction rules to run on databases, which can come in useful if we have to process large sets of training or evaluation data.

Web Block Classification. It does not seem feasible to solve the block classification problem through a set of logic-based rules, as it is often the case that there are many potential features which can be used to characterise a specific block type. However, only a few play a major role in uniquely distinguishing this block type from all others. Some of these features are continuous (e.g. the block's width and height), and it can be difficult for a human to manually specify accurate threshold boundaries. Hence, for BER$_y$L, we decided to use a machine learning (ML) approach to Web block classification.

Comparative Analysis of Machine Learning Techniques. The current version of the BER$_y$L Web block classification system uses the C4.5 Decision Tree as the classification model, as it allows us to check how the features are used in rules that guide the overall classification, which is not the case with other ML classification models such as SVMs. Also, the rules generated by the C4.5 Decision Tree can be easily translated into Datalog rules, which can become useful if we decide to run the ML classification stage of BER$_y$L in Datalog, same as the feature extraction stage.

4 Evaluation

In Fig. 5 we present the evaluation results for five domain-independent block classifiers (headers, footers, sidebars, navigation menus, and next links) obtained on 500 randomly-selected pages from four different domains (Real Estate, Used

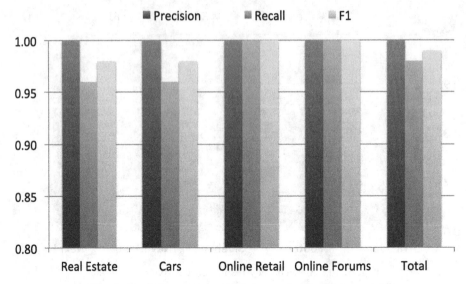

Fig. 6. Precision and recall of the pagination link model

Cars, Online Retail, Blogs and Forums). For each block type and domain, the pages have been selected randomly from a listings page (such as yell.com) or from a Google search. The latter favours popular websites, but that should not affect the results presented here. For all these classifiers we have achieved good precision, recall, and F_1 scores. All the classifiers have precision and recall scores above 80% apart from the Sidebar classifier, which has a precision score of just below 80%. This can be explained by the fact that a sidebar usually does not have obvious clues, such as the NEXT annotation for non-numeric next links, and therefore it is harder to distinguish True Positives from False Positives and False Negatives. Note that the Next Link and Header classifiers achieve a perfect precision of 100%. This can be explained by the fact that we use a highly tailored feature set that includes features specific to the Next Link block, and that headers are very distinct from other blocks, and our fairly simple set of six features used for the classification of this block is sufficient to achieve a perfect separation between header and non-header DOM nodes.

Accuracy of Pagination Link Classification. We put a special emphasis on the Pagination Link classifier, due to its importance to the DIADEM system, and hence have performed a more detailed evaluation for this block type than for the other three. We present detailed evaluation results for the Pagination Link classifier in Fig. 6, which illustrates that in all four domains our approach achieves 100% precision, and recall is never below 96%. This high accuracy means that our approach can be used to crawl or otherwise navigate paginated websites with a very low risk of missing information or retrieving unrelated Web pages. Numeric pagination links are generally harder to classify than non-numeric ones, due to their greater variety and the larger set of candidates. Though precision

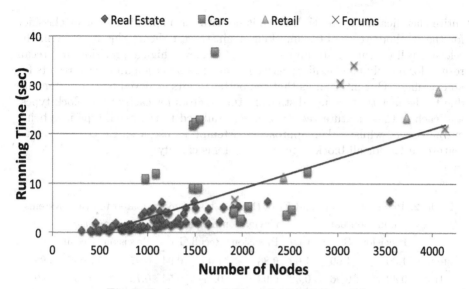

Fig. 7. Performance of the pagination link model

is 100% for both cases, recall is on average slightly lower for numeric pagination links (98% vs. 99%) and in some domains quite notable (e.g. Real Estate with 96% vs. 99%).

Performance Results. The speed of feature extraction is crucial for the scalability of our approach to allow the crawling of entire websites. As discussed above, the use of visual features by itself imposes a certain penalty, as a page needs to be rendered for those features to be computed. We present the performance results of BER$_y$L in Fig. 7, which shows, for the example of the Pagination Link classifier, that the performance is highly correlated to page size, with most reasonably sized pages being processed in well below 10 seconds (including page fetch and rendering). It is interesting to observe that the domains for which we use Google to generate the corpus, and for which the corpus is thus biased towards popular websites, seem to require more time than the Real Estate domain, for which the corpus is randomly picked from yell.com.

Comparison to Other Approaches. We compare the precision, recall, and F$_1$ results achieved by the BER$_y$L system with the results of other systems that were covered in Sect. 2. Some of the papers analysed there do not present the accuracy results, so we omit them from this comparison, and for others we compare to the approaches, which cover the block types also covered by the BER$_y$L system, and if there is no overlap between the block types covered, we compare on the average accuracy results for all classifiers in the system. As a lot of block types covered by our system are not covered by other approaches, we compare the precision, recall, and F$_1$ metrics over two block types where there is an overlap (navigation menus and pagination bars), and the average precision, recall, and F$_1$ metrics for entire classification systems. If there is no data for a specific metric or classifier for the method we are benchmarking against, we indicate this by N/A in the relevant cell of Table 2. In all cases BER$_y$L achieves higher precision and recall results for individual classifiers, as well as average precision and recall results for all classifiers. Our intuition is that the better results achieved by our system are due to the fact that we use distinct feature vectors for each of the block types, and each of these feature vectors is highly tailored to the block type it is being extracted on, whilst other approaches attempt to use a single coarse-grained feature vector for all block types their systems classify.

Table 2. Precision, recall, and F$_1$ of the BER$_y$L system compared to other systems

	Pagination bars			Navigation menus			System average		
	Precision	Recall	F$_1$	Precision	Recall	F$_1$	Precision	Recall	F$_1$
BER$_y$L	1.00	1.00	1.00	0.88	1.00	0.94	0.97	1.00	0.98
[15]	0.42	0.96	0.58	0.98	0.37	0.54	0.73	0.65	0.69
[24]	N/A	N/A	N/A	N/A	N/A	0.88	N/A	N/A	0.75
[20]	N/A	N/A	0.82	N/A	N/A	0.82	N/A	N/A	0.52
[12]	N/A	N/A	N/A	N/A	N/A	N/A	0.77	0.71	0.74
[17]	N/A	N/A	N/A	N/A	N/A	N/A	0.75	0.66	0.70

5   Conclusion

We propose a Web block classification system, BER$_y$L, which utilises global feature and template repositories for providing substantial improvement in the manual effort of defining new features and improving the performance of feature extraction.

We aim to pursue multiple avenues for future research, in particular (**1**) further exploration of how domain knowledge can improve block classification and the differentiation between domain-independent and domain-dependent classifiers, (**2**) exploration of holistic approach to Web block classification, which imposes constraints between different block types, e.g. that a footer cannot be above a header, (**3**) exploration of the α-accuracy problem, which allows the system to find an optimal balance between the accuracy and performance expectations set by the user, and (**4**) automatic learning of features for individual block classifiers.

References

1. Abiteboul, S., Hull, R., Vianu, V.: Foundations of Databases. Addison-Wesley Longman Publishing Co. Inc., Boston (1995)

2. Baumgartner, R., Flesca, S., Gottlob, G.: Visual web information extraction with Lixto. In: VLDB (2001)
3. Baluja, S.: Browsing on small screens: recasting web-page segmentation into an efficient machine learning framework. In: WWW 2006 (2006)
4. Burget, R., Rudolfova, I.: Web page element classification based on visual features. In: 2009 First Asia Conference on Intelligent Information and Database Systems (2009)
5. Cai, D., Yu, S., Wen, J., Ma, W.: Block-based web search. In: SIGIR 2004, 25–29 July 2004 (2004)
6. Cai, D., He, X., Wen, J., Ma, W.: Block-level link analysis. In: SIGIR 2004, 25–29 July 2004 (2004)
7. Cao, Y., Niu, Z., Dai, L., Zhao, Y.: Extraction of informative blocks from web pages. In: ALPIT 2008 (2008)
8. Chen, J., Zhou, B., Shi, J., Zhang, H., Fengwu, Q.: Function-based object model towards website adaptation. In: WWW 2010, 1–5 May 2010 (2010)
9. de Moor, O., Gottlob, G., Furche, T., Sellers, A. (eds.): Datalog 2.0 2010. LNCS, vol. 6702. Springer, Heidelberg (2011). https://doi.org/10.1007/978-3-642-24206-9
10. Furche, T., Grasso, G., Kravchenko, A., Schallhart, C.: Turn the page: automated traversal of paginated websites. In: Brambilla, M., Tokuda, T., Tolksdorf, R. (eds.) ICWE 2012. LNCS, vol. 7387, pp. 332–346. Springer, Heidelberg (2012). https://doi.org/10.1007/978-3-642-31753-8_27
11. Furche, T., et al.: DIADEM: domain-centric, intelligent, automated data extraction methodology. In: WWW 2012 (2012)
12. Goel, A., Michelson, M., Knoblock, C.A.: Harvesting maps on the web. Int. J. Doc. Anal. Recognit. **14**(4), 349 (2011)
13. Gottlob, G., Orsi, G., Pieris, A., Simkus, M.: Datalog and its extensions for semantic web databases. In: Eiter, T., Krennwallner, T. (eds.) Reasoning Web 2012. LNCS, vol. 7487, pp. 54–77. Springer, Heidelberg (2012). https://doi.org/10.1007/978-3-642-33158-9_2
14. Gupta, S., Kaiser, G., Neistadt, D., Grimm, P.: DOM-based content extraction of HTML documents. In: WWW 2003, 20–24 May 2003 (2003)
15. Kang, J., Choi, J.: Block classification of a web page by using a combination of multiple classifiers. In: Fourth International Conference on Networked Computing and Advanced Information Management, 2–4 September 2008 (2008)
16. Kang, J., Choi, J.: Recognising informative web page blocks using visual segmentation for efficient information extraction. J. Univ. Comput. Sci. **14**(11), 1893 (2008)
17. Keller, M., Hartenstein, H.: GRABEX: a graph-based method for web site block classification and its application on mining breadcrumb trails. In: 2013 IEEE/WIC/ACM International Conferences on Web Intelligence (WI) and Intelligent Agent Technology (IAT) (2013)
18. Kordomatis, I., Herzog, C., Fayzrakhmanov, R.R., Krüpl-Sypien, B., Holzinger, W., Baumgartner, R.: Web object identification for web automation and meta-search. In: WIMS 2013 (2012)
19. Krüpl-Sypien, B., Fayzrakhmanov, R.R., Holzinger, W., Panzenböck, M., Baumgartner, R.: A versatile model for web page representation, information extraction and content re-packaging. In: DocEng 2011, 19–22 September 2011 (2011)
20. Lee, C.H., Kan, M., Lai, S.: Stylistic and lexical co-training for web block classification. In: WIDM 2004, 12–13 November 2004 (2004)
21. Li, C., Dong, J., Chen, J.: Extraction of informative blocks from web pages based on VIPS. J. Comput. Inf. Syst. **6**(1), 271 (2010)

22. Liu, W., Meng, X.: VIDE: a vision-based approach for deep web data extraction. IEEE Trans. Knowl. Data Engineering **22**(3), 447 (2010)
23. Luo, P., Lin, F., Xiong, Y., Zhao, Y., Shi, Z.: Towards combining web classification and web information extraction: a case study. In: KDD 2009, 28 June–1 July (2009)
24. Maekawa, T., Hara, T., Nishio, S.: Image classification for mobile web browsing. In: WWW 2006, 23–26 May (2006)
25. Romero, R., Berger, A.: Automatic partitioning of web pages using clustering. In: Brewster, S., Dunlop, M. (eds.) Mobile HCI 2004. LNCS, vol. 3160, pp. 388–393. Springer, Heidelberg (2004). https://doi.org/10.1007/978-3-540-28637-0_43
26. Song, R., Liu, H., Wen, J., Ma, W.: Learning block importance models for web pages. In: WWW 2004, 17–22 May (2004)
27. Vadrevu, S., Velipasaoglu, E.: Identifying primary content from web page and its application to web search ranking. In: WWW 2011 (2011)
28. Wang, J., et al.: Can we learn a template-independent wrapper for news article extraction from a single training site? In: KDD 2009, 28 June–1 July (2009)
29. Wu, C., Zeng, G., Xu, G.: A web page segmentation algorithm for extracting product information. In: Proceedings of the 2006 IEEE International Conference on Information Acquisition, 20–23 August 2006 (2006)
30. Xiang, P., Yang, X., Shi, Y.: Effective page segmentation combining pattern analysis and visual separators for browsing on small screens. In: Proceedings of the 2006 IEEE/WIC/ACM International Conference on Web Intelligence (2006)
31. Xiang, P., Yang, X., Shi, Y.: Web page segmentation based on gestalt theory. In: 2007 IEEE International Conference on Multimedia and Expo (2007)
32. Yang, X., Shi, Y.: Learning web block functions using roles of images. In: Third International Conference on Pervasive Computing and Applications, 6–8 October 2008 (2008)
33. Yi, L., Liu, B., Li, X.: Eliminating noisy information in web pages for data mining. In: SIGKDD 2003, 24–27 August 2003 (2003)
34. Yu, S., Cai, D., Wen, J., Ma, W.: Improving pseudo-relevance feedback in web information retrieval using web page segmentation. In: WWW 2003, 20–24 May 2003 (2003)
35. Zheng, S., Song, R., Wen, J., Giles, C.L.: Efficient record-level wrapper induction. In: CIKM 2009, 2–6 November 2009 (2009)
36. Zhu, J., Nie, Z., Wen, J., Zhang, B., Ma, W.: Simultaneous record detection and attribute labeling in web data extraction. In: KDD 2006, 20–23 August 2006 (2006)

The Refutation of Amdahl's Law
and Its Variants

F. Dévai[1,2]([⊠])(iD)

[1] London South Bank University, London, UK
fdevai@acm.org
[2] Hungarian Academy of Sciences, Budapest, Hungary

Abstract. Amdahl's law, imposing a restriction on the speedup achievable by a multiple number of processors, based on the concept of sequential and parallelizable fractions of computations, has been used to justify, among others, asymmetric chip multiprocessor architectures and concerns of "dark silicon". This paper demonstrates flaws in Amdahl's law that (i) in theory no inherently sequential fractions of computations exist (ii) sequential fractions appearing in practice are different from parallelizable fractions and usually have different growth rates of time requirements and that (iii) the time requirement of sequential fractions can be proportional to the number of processors. However, mathematical analyses are also provided to demonstrate that sequential fractions have negligible effect on speedup if the growth rate of the time requirement of the parallelizable fraction is higher than that of the sequential fraction. Examples are given that Amdahl's law and its variants fail to represent limits to parallel computation. In particular, Gustafson's law, claimed to be a refutation of Amdahl's law by some authors, is shown to contradict established theoretical results. We can conclude that no simple formula or law governing concurrency exists.

Keywords: Amdahl's law · Gustafson's law
Sequential and parallelizable workload · Growth rates
Asymmetric chip multiprocessor architectures
Graphics processing units · Hidden-surface removal
Inherently sequential computations · P-complete problems
Data-parallel computing

1 Introduction

The computing literature about parallel and distributed computing can roughly be divided in two categories: publications relying on Amdahl's law [4] and its variants, as well as publications about designing and implementing multiprocessor algorithms. Amdahl's law and its variants, including Gustafson's law [33], provide little help with issues the second group of publications are dealing with, therefore hardly ever mentioned by that group of publications. One

© Springer-Verlag GmbH Germany, part of Springer Nature 2018
M. L. Gavrilova and C. J. K. Tan (Eds.): Trans. on Comput. Sci. XXXIII,
LNCS 10990, pp. 79–96, 2018.
https://doi.org/10.1007/978-3-662-58039-4_5

of the few exceptions is Preparata [60] asking in an 1995 invited presentation if Amdahl's law should be repealed. However, on 17 January 2018, according to Google Scholar, Amdahl's paper [4] has been cited by 4977, and Gustafson's paper [33] by 1774 publications, including several published in 2017.

Both Amdahl's law and Gustafson's law attempt to predict the speedup of computation time achievable by a multiple number of processors. Amdahl [4] places a pessimistic upper bound on the multiprocessor approach by concluding that the achievable speedup is a constant of five to seven, even if the number of processors approaches infinity.

On the other hand, Gustafson's law is overly optimistic. Gustafson [33] states that the achievable speedup is proportional to the number of processors. Having two drastically different versions of a scientific law, up for choice by the advocates, is awkward. For example, if a re-evaluated version of Ohm's law predicted more electric current than voltage divided by resistance, it would not be credible. Similarly, one of Amdahl's law and Gustafson's law cannot be correct.

This paper demonstrates that both Amdahl's law and its variants, including Gustafson's law, are fundamentally wrong. Some results from the second group of publications will be used, but interestingly, none of the publications used ever mention Amdahl's law and its variants.

Sections 2 and 3, respectively, discuss Amdahl's law and Gustafson's law in more detail. Section 4 reviews some of the most frequently cited and most recent publications advocating Amdahl's law and its variants. Section 5 reviews earlier attempts to refute Amdahl's law and summarises arguments that no substantial impediments to parallel computation exist. Section 6 summarises models of computation. Section 7 reviews data-parallel computing. Section 8 points out several flaws in Amdahl's law. Section 9 demonstrates that Gustafson's law contradicts established theoretical results. Section 10 reviews the possibilities for the development of fast parallel algorithms and the known limits to parallel computation. Section 11 discusses the contributions of this paper. Finally Sect. 12 concludes that no simple formula or law governing concurrency exists and that experimental results should be interpreted with theoretical results in mind.

2   Amdahl's Law

Amdahl's paper [4] published in 1967, criticizes *"prophets"* voicing *"the contention that the organization of a single computer has reached its limits and that truly significant advances can be made only by interconnection of a multiplicity of computers"*. Amdahl argues that *"the fraction of the computational load ... associated with data management housekeeping ... accounts for 40% of the executed instructions"*.

He goes on that this fraction *"might be reduced by a factor of two"* but it is highly unlikely that *"it could be reduced by a factor of three"* and that *"this overhead appears to be sequential so that it is unlikely to be amenable to parallel processing techniques"*. He concludes that this overhead places *"an upper limit on throughput of five to seven times the sequential processing rate"*.

Amdahl aims to provide an upper limit on speedup and therefore assumes that, apart from the sequential fraction, the remaining computations are perfectly parallelizable. Let t_1 be the time taken by one processor solving a computational problem and t_p be the time taken by p processors solving the same problem. Finally let us denote the supposed inherently sequential fraction of instructions by f. Then, according to Amdahl, $t_p = t_1(f + (1 - f)/p)$ and the speedup obtainable by p processors can be expressed as

$$\frac{t_1}{t_p} = \frac{1}{f + (1 - f)/p}. \tag{1}$$

Indeed if we substitute $f = 0.4/2$ and $f = 0.4/3$, and assume that p approaches infinity, we get 5 and 7.5 respectively for speedup. Formula (1) is referred to as Amdahl's law by his followers (though Amdahl did not present the formula).

3   Gustafson's Law

Gustafson observed experimental results contradicting Amdahl's law. He reports that a research group at Sandia National Laboratories, US, achieved speedup 1016 to 1021 on a 1024-processor hypercube [33]. Gustafson claims that the amount of work that can be done in parallel grows with the problem size and also with the number of processors, while serial bottlenecks, including input-output, do not grow with the problem size. The reason given for this by Gustafson is that if, e.g., we double the number of processors, we can also double the number of degrees of freedom in a physical simulation [33].

Using high-performance computing facilities, such as a 1024-processor hypercube, it is the parallel time t_p on p processors what the user can readily observe. Gustafson argues that a single processor solving the same problem would spend ft_p time on the sequential, and $(1 - f)pt_p$ time on the parallelizable part of the problem, therefore $t_1 = ft_p + (1 - f)pt_p$, from which the speedup achievable by p processors, known as Gustafson's law, is

$$\frac{t_1}{t_p} = f + (1 - f)p, \tag{2}$$

where f is the same "inherently sequential" fraction of instructions as in the case of Amdahl's law. As Gustafson's law is based on the same concepts as the bases of Amdahl's law, it is a variant, rather than a refutation of Amdahl's law. Indeed, the title of Gustafson's paper [33] is *"Reevaluating Amdahl's Law"*.

Several authors, e.g., Denning and Lewis [17] and Sun and Chen [65], accept Gustafson's law as the refutation of Amdahl's law. However, Amdahl's law is still more widely accepted than Gustafson's law.

One reason behind this could be Gustafson's claim [33] that, while the parallelizable work grows with the problem size, Amdahl's law is still valid if the problem size is fixed, even though Amdahl never stipulated this requirement.

Gustafson's claim led to the acceptance of Amdahl's law as the *fixed-size*, while Gustafson's law as the *fixed-time speedup law* [12,41,50]. However, the concept of fixed-size speedup is not well defined: fixing the problem size at different points would result in different speedups.

4 Amdahl's and Gustafson's Followers

Gustafson does not take sequential input-output requirements proportional to input and output sizes into account. Patterson et al. [57], in an attempt to defeat Amdahl's law, propose disk arrays to reduce input-output requirements.

In order to mitigate the effects of Amdahl's law, Annavaram et al. [6] make a case for varying the amount of energy expended to instructions according to the amount of available parallelism. Power consumption is the product of energy per instruction (EPI) and instructions per second (IPS). Annavaram et al. [6] propose that during phases of limited parallelism (low IPS) a chip multiprocessor should spend more EPI, and during phases of higher parallelism (high IPS) it should spend less EPI.

Borkar [8], without explicitly citing Amdahl's paper, presents (1) and argues that, due to the limitation by Amdahl's law, it will be difficult to exploit the performance enhancements of many-core architectures. However, he adds that multiple applications running simultaneously mitigate the effect of Amdahl's law and therefore could benefit from many-core architectures.

Many authors [37,45,64,71] propose asymmetric (or heterogeneous) chip multiprocessor architectures consisting of at least one large, high-performance core and several small, low-performance cores. Serial program portions would execute on a large core to reduce the performance impact of the serial bottleneck imposed by Amdahl's law. Some of these proposals are summarised below.

Kumar et al. [45], relying on Annavaram et al. [6], claim that asymmetric chip multiprocessor architectures present unique opportunities for mitigating the adverse effect of Amdahl's law on the achievable speedup. Hill and Marty [37] argue that robust, general-purpose multicore designs should operate under Amdahl's more pessimistic, rather than Gustafson's optimistic assumptions. They assume that architects have techniques for using the resources of multiple cores to create a single core with greater sequential performance. However, Hill and Marty [37] conclude that the future of scalable multicore processors is questionable.

Woo and Lee [71] attempt to extend Amdahl's law in order to take power and energy into account, independently of Annavaram et al. [6]. They suggest many small, energy-efficient cores integrated with a large powerful core.

Suleman et al. [64] propose a technique, called Accelerated Critical Sections (ACS), leveraging the high-performance core(s) of asymmetric chip multiprocessors to accelerate the execution of critical sections. When a small core encounters a critical section, it requests a large one to execute that critical section. The large core acquires the lock, executes the critical section and notifies the requesting small core when the critical section is complete.

As Suleman et al. [64] properly note, ACS have disadvantages. In order to execute critical sections, the large core may require some private data from the small core, e.g., input parameters on a stack. Such data is transferred from the cache of the small core via the regular cache coherence mechanism. These transfers increase cache misses.

Executing critical sections exclusively on a large core can have another disadvantage. Paraphrasing Suleman et al. [64], multi-threaded applications often try to improve concurrency by using data synchronization at a fine granularity, i.e., having multiple critical sections, each guarding a disjoint set of the shared data (e.g., a separate lock for each element of an array). Then executing all critical sections on the large core can lead to false serialization of different, disjoint critical sections that otherwise could have been executed in parallel.

Sun and Chen [65] study symmetric multicore architectures. As noted above, they accept Gustafson's law and criticize Hill and Marty's pessimistic view [37]. They conclude that asymmetric multicore architectures are much more complex and therefore are worth exploring only if their symmetric counterparts cannot deliver satisfactory performance.

However, according to Borkar and Chien [9], there is another reason why multicore architectures can be asymmetric and heterogeneous: Cores in close proximity can be interconnected into clusters by using busses that are energy efficient for data movement over short distances. Then the clusters can be connected by using packet- or circuit-switched networks.

Eyerman and Eeckhout [23] propose to augment Amdahl's law with the notion of critical sections. They claim that it leads to a new fundamental law, asserting that parallel performance is not only limited by the sequential fraction (according to Amdahl's law) but also limited by critical sections.

Juurlink and Meenderinck [39] consider Amdahl's law harmful. In an attempt to reconcile the differences between Amdahl's law and Gustafson's law, they offer a new law assuming that the parallelizable work grows as a *sublinear* function of p, the number of processors. In fact, the non-parallelizable work can grow as a *linear* function of p, as we demonstrate in Sect. 8.

Esmaeilzadeh et al. [22] predict that as the number of cores increases, power constraints may prevent powering up all the cores at full speed. According to their models—based on Amdahl's law—the fraction of chips remaining "dark" could be as much as 50% within three process generations.

Yavits et al. [72] consider asymmetric chip multiprocessor architectures, with a general-purpose core responsible for executing the sequential fraction and a number of other cores for executing the parallelizable fraction of the code. This arrangement incurs data exchange between the sequential and the parallel cores and also a data exchange among the parallel cores. Based on these observations, the authors suggest a modification of Amdahl's law, offering a pessimistic conclusion that a highly parallelizable yet highly synchronization- and connectivity-intensive workload might be more efficiently processed by a sequential core rather than a parallel multicore.

Morad et al. [53] propose a shared memory to speed up communication between parallel and sequential cores. Yavits et al. [73] study the thermal effects of three-dimensional integration on the performance and scalability of chip multiprocessors from the perspective of Amdahl's law.

Mittal [52] provides a comprehensive survey on asymmetric multicore processors, mainly justified by Amdahl's law, emphasising the challenges these architectures face, including the optimization of both sequential and parallel performance and thread migration overheads. The latter can take millions of cycles, based on independent experimental results taken from the literature.

Patterson and Hennessy [55] and Hennessy and Patterson [34, pp. 46–47] generalize Amdahl's law for any performance enhancement by defining speedup as the execution time for a particular task without using the enhancement divided by the execution time for the same task using the enhancement. If the enhancement cannot affect, say, one third of the execution time of the task, the speedup can only be at most threefold. This is called *the law of diminishing returns*, which in itself is correct, regardless of the validity of Amdahl's law [56]. For the rejection of Gustafson's law, Hennessy and Patterson [34] refer to Amdahl's paper [4] without explicitly citing Gustafson's paper [33].

Shavit [63] presents Amdahl's law (without explicitly citing Amdahl's paper) and suggests that it is worthwhile to invest efforts to derive as much parallelism as possible from parts of programs dealing with inter-thread interaction and coordination in order to get highly parallel, concurrent data structures. This is an important research direction, irrespective of the validity of Amdahl's law.

The textbook by Herlihy and Shavit gives an example of a 10-processor machine and a program with a 90% parallelizable fraction [36, p. 14] and states that according to Amdahl's law (1) the speedup is merely 5.26. However, the authors also state that the major focus of their book is to help understanding the tools and techniques that allow software developers effectively preparing the parts of the program code dealing with coordination and synchronization. This approach makes the book useful, regardless of the validity of Amdahl's law.

5   No Substantial Impediments to Parallelism Exist

In 1975 Valiant [66] proved that, using p processors doing binary comparisons, speedup proportional to $p/\log\log p$ can be achieved for problems of finding the maximum, sorting, and merging a pair of sorted lists, assuming $p \leq n$, where n the size of the input set. These results assume that the input data are already in memory, while Amdahl's law also involves *"data management housekeeping"* which includes bringing data into memory. Therefore these results cannot be regarded as the refutations of Amdahl's law. However, Valiant's results, generalized to other models of parallel computation [24,25,42], indicate that, as Valiant noted in 1990, *"no substantial impediments to general-purpose parallel computation"* exist [67] though there are limits, as shown in Sect. 10 below.

In 1995 Preparata [60] stated that research on algorithms has shown that most problems are parallelizable and that, in a realistic computational model fully accounting for the finiteness of the speed of light, uniprocessors incur a slowdown not only due to loss of parallelism but also due to loss of locality.

Preparata [60] assumed that Amdahl's law may only apply to P-complete problems. In Sect. 10 we demonstrate that even some P-complete problems are parallelizable, such that the product of the parallel running time and the number of processors used matches the sequential lower bound for the problem, and therefore breaking Amdahl's law.

Although Amdahl's law has already been classified as a "folk theorem" [1, 44] and claimed to be refuted for a special case of *superlinear speedup*, when the amount of input data keeps increasing during computation [1,47] it is still widely accepted. Superlinear speedup is nicely illustrated by Luccio and Pagli's [47] snow-shoveler metaphor: In a winter morning a man shovels snow from his driveway, while the snow falls. On a neighbouring driveway three men do the same job. As a smaller amount of snow falls while they work, they clean the driveway in less than one third of their neighbour's time.

In 2007 Paul and Meyer [58] stated that Amdahl's law is *"one of the few, fundamental laws in computing"* but went on that it can fail when applied to single-chip heterogeneous (asymmetric) multiprocessor designs. However, as we have already seen in Sect. 4, even after Paul and Meyer's publication [58] much research continued on the mitigation of the adverse effect of Amdahl's law on asymmetric chip multiprocessors.

Some of these researchers [23,63,64,72] observed that, in practice, parallel performance is not only limited by a sequential fraction, as Amdahl's law claims, but also limited by communication, synchronization and concurrent objects. In Sect. 8 we demonstrate that the time requirement of these can be proportional to the number of processors.

Amdahl's and Gustafson's followers never gave examples of "inherently sequential" computations and not even specific examples when sequential computations appear in concurrent systems in practice. The theory community never provided refutations, merely ignores Amdahl's law and its variants.

In Sect. 8 we demonstrate that inherently sequential computations do not exist in theory. We also demonstrate that Amdahl's law is broken in practice by using three practical examples of sequential operations (such as reading input, sequential concurrent objects [21] and Lamport's bakery algorithm [46]). Then in Sect. 9 we show that Gustafson's law contradicts established theoretical results. But before that we need to consider models of computation and data-parallel computing.

6   The Parallel Random-Access Machine

The most widely accepted model of parallel computation is the parallel random-access machine (PRAM) [29]. A PRAM consists of a set of processors, which are random-access machines (RAMs) [15], attached to a single shared memory, called the global memory. The processors communicate through the global memory: each processor can write a value into a location (word) of the memory and another can read this value. All operations are assumed to take constant time. Deterministic PRAMs can accept, in polynomial time, exactly the same sets accepted by polynomial tape-bounded Turing machines [29].

Though the PRAM has been introduced as a theoretical model, a hierarchy of fast cache memories can be used to alleviate the difficulties of implementing shared memory. It is advocated that, due to this memory hierarchy, PRAM algorithms can directly be implemented on chip multiprocessors [27,28,31,68].

Several variants of the PRAM model have been used that differ in whether or not they permit concurrent access to a particular memory location and in how to resolve conflicts during these concurrent accesses. These variants are the exclusive read, exclusive write (EREW), concurrent read, exclusive write (CREW) and the concurrent read, concurrent write (CRCW) models.

The most widely used is the CREW variant, while the most powerful one is the CRCW variant. The CRCW variant has three subcategories [32,43]. In the case of the COMMON CRCW subcategory, concurrent writes only succeed if all processors write the same value. In the PRIORITY CRCW PRAM only the write by the lowest numbered contending processor succeeds, while in the ARBITRARY subcategory any one of the writes succeeds.

7 Data-Parallel Computing

It would be tempting to believe that Gustafson's law applies to data-parallel computing. For example, Denning and Lewis claim that *"Gustafson's Law models data-parallel computing"* [17] and that *"Gustafson's Law gives bounds on data-parallel processing"* [19, Authors Respond]. On the other hand, even publications on data-parallel computing advocate Amdahl's law [11,17].

Hillis and Steele [38] introduced a programming style by describing a series of algorithms, called data-parallel algorithms, for fine-grained parallel computers with general communications. The parallelism of these algorithms results from simultaneous operations across large sets of data, rather than from multiple threads of control. Data parallelism is usually implemented by parallel loops. For example, the maximum of n numbers can be found as follows.

Let A be an n-element linear array containing the n input numbers and let M be another n-element linear array that can store the Boolean values TRUE and FALSE. Using n^2 COMMON CRCW PRAM processors, we can execute the following three parallel loops [43].

> **for** $i \leftarrow 1$ **to** n **in parallel do** $M[i] \leftarrow$ TRUE;
> **for** $i \leftarrow 1$ **to** n **and** $j \leftarrow 1$ **to** n **in parallel do**
>   **if** $A[i] < A[j]$ **then** $M[i] \leftarrow$ FALSE;
> **for** $i \leftarrow 1$ **to** n **in parallel do if** $M[i] =$ TRUE **then** $max \leftarrow A[i]$

This code finds the maximum in constant time, but for any PRAM without simultaneous writes, there exists an $\Omega(\log n)$ lower bound allowing even infinitely many processors [14]. This demonstrates that data-parallel computing is a programming style [38], rather than an alternative model of computation.

Several environments have been developed for data-parallel computing [2,16,54,70]. The most spectacular one is MapReduce [16] including a user interface and an underlying runtime system. The computational problem is specified in terms of a *map* and a *reduce* function. Then the runtime system automatically parallelizes the computation, scheduling inter-processor communication and handling machine failures.

Hadoop [70] is the open-source implementation of MapReduce, while CUDA (an acronym for Compute Unified Device Architecture) is a data-parallel computing platform and application-programming interface for graphics processing units (GPUs) and GPU clusters [54].

As noted above, data-parallel computing is not an alternative model of computation. For example, Karloff, Suri, and Vassilvitskii [40] proved that a large class of PRAM algorithms can efficiently be simulated via MapReduce.

8   The Refutation of Amdahl's Law

One of the most spectacular success stories of the computing industry is the parallelization of the hidden-surface problem by using GPUs. The hidden-surface problem is stated as follows.

Given a set S of pairwise disjoint, opaque and planar simple polygons possibly with holes and with a total of n edges in three-dimensional space; find each region ρ of each polygon in S, such that all points of ρ are visible from a viewpoint u, where $u = (0, 0, \infty)$.

The best possible hidden-surface algorithms take time proportional to n^2 in the worst-case [18,51]. In GPUs the z-buffer [5], also called the depth-buffer [48] approximation algorithm is used. The input polygons are decomposed into triangles. The hidden-surface image of 10 random triangles in three-dimensional space is given in Fig. 1.

Based on a result by Alikoski [3,59] we can prove that the z-buffer algorithm also takes time proportional to n^2 even on the average.

Theorem 1 (Alikoski, 1938). *Taking three points uniformly, independently, at random from the unit square, the expected area of the triangle determined by the three points is $11/144$.*

Corollary 1. *The expected time for the z-buffer algorithm to display n triangles on r^2 picture elements is $\Theta(nr^2)$ assuming that all triangles are equally likely.*

It is reasonable to assume at least as many picture elements as triangles, i.e., $r^2 \geq n$, therefore it is a conservative estimate that the expected running time of the z-buffer algorithm is proportional to n^2.

The input set S is usually too large to fit in main memory, therefore we need to read the relevant parts from a disk or a solid-state drive. Ignoring disk seek time, reading the data sequentially takes time proportional to n. Although the output can be as large as $O(n^2)$, it is perfectly parallelizable; each processor writes its result into a memory block, called the frame buffer. Then the running time for a single processor is

$$t_1(n) = an + bn^2$$

for some positive constants a and b.

Fig. 1. A hidden-surface image

For some small n, we can observe sequential reading times of, e.g., one third or one fourth of $t_1(n)$. Substituting $f = 1/3$ or $1/4$ into (1) in Sect. 2, we would get upper limits of 3 or 4 for speedup, while GPUs successfully use hundreds or thousands of processors.

To resolve the contradiction, we should go back to the original definition of speedup: $t_1(n)$ divided by $t_p(n)$. According to Amdahl (Sect. 2) for the development of an upper bound we assume that, apart from the sequential fraction, the remaining computations are perfectly parallelizable. Therefore

$$t_p(n) = an + \frac{bn^2}{p},$$

from which it follows that

$$\frac{t_1(n)}{t_p(n)} = \frac{a + bn}{a + bn/p}.$$

If p approaches infinity, we get the upper limit

$$\frac{t_1(n)}{t_p(n)} = 1 + \frac{b}{a}n \qquad (3)$$

that grows linearly with n, rather than *"an upper limit"* of *"five to seven"*.

However, it is a more realistic assumption that p is proportional to n. If we substitute $n = cp$, for some positive constant c, we get

$$\frac{t_1(n)}{t_p(n)} = \frac{a + bcp}{a + bc},$$

a speedup that grows linearly with p.

Even though Amdahl's law is overly pessimistic, it does not take into account that increasing the number of processors can have adverse effect on speedup. Ellen, Hendler and Shavit [21,26] show that performing p operations, each by a different process on objects with nonblocking linearizable implementations, takes p times the time of a single operation. Also the implementation of Lamport's bakery algorithm [46] for p processes takes time proportional to p.

Assuming that the parallelizable fraction has a quadratic growth rate as before and that the number of processes is the same as the number of processors, using the notation as above we have

$$\frac{t_1(n)}{t_p(n)} = \frac{ap + bn^2}{ap + bn^2/p}.$$

Substituting $n = cp$, we get

$$\frac{t_1(n)}{t_p(n)} = \frac{a + bc^2 p}{a + bc^2},$$

a speedup that still grows linearly with p.

It should be noted that the above is not a new law that any problem with a linear sequential fraction and a quadratic parallelizable fraction would have a speedup proportional to the number of processors. It merely indicates that the speedup is heavily dependent on the growth rates of the fractions of the code executed sequentially and in parallel. If the growth rate of the fraction executed sequentially is the same or higher than that of the fraction executed in parallel, there would indeed be a small upper limit for speedup.

To examine if this is a possibility, we need to see if any fraction of instructions executed sequentially is indeed *"unlikely to be amenable to parallel processing techniques"*. A result by Dymond and Tompa suggests that this is not the case. Dymond and Tompa [20] proved that every deterministic Turing machine running in time t can be simulated by a CREW PRAM in time $O(t^{1/2})$.

The linear tapes of the Turing machine forbid random access of individual tape cells. Dymond and Tompa did not clarify if the more flexible storage structure of the PRAM, the parallelism, or the combination of both realizes such a spectacular speedup. Mak [49] has demonstrated that parallelism alone suffices to achieve an almost quadratic speedup. From here it follows that no inherently sequential computations exist in theory.

All the above demonstrate that Amdahl's law is fundamentally wrong. The usual objections to any criticism of Amdahl's law are that the critics violate Amdahl's assumptions [10,35]. Amdahl's fundamental assumption is that the fraction of the computational load remaining sequential determines the achievable speedup. This assumption has not been violated above.

The fact that the parallelizable fraction may grow with the problem size has already been experimentally oserved, among others by Gustafson (Sect. 3) and a more general reason for it has been given above.

9   The Refutation of Gustafson's Law

Gustafson's law contradicts not only the results by Dymond and Tompa and Mak, but also other established theoretical results known before Gustafson's publication [33] in 1988, as already noted [19] and shown below.

The sequential running time for finding the maximum of n integers, $t_1(n) \leq cn$, accounting for $n - 1$ comparisons and, in the worst case, $n - 1$ assignments, where c is a positive constant. Based on Gustafson's law (2) the time to find the maximum of n integers by using p processors would be

$$t_p(n) \leq \frac{cn}{f + (1 - f)p},$$

which is bounded by a constant for any $0 < f < 1$ and any p proportional to n, and approaches 0 for any fixed n if p approaches infinity.

However, as noted above, in 1982 Cook and Dwork [14] provided an $\Omega(\log n)$ lower bound for finding the maximum of n integers allowing infinitely many processors of any PRAM without simultaneous writes.

In 1985 Fich et al. [24] proved an $\Omega(\log \log n)$ lower bound for the same problem for $p < \binom{n}{2}$ under the PRIORITY CRCW PRAM. The PRIORITY model is the strongest PRAM model allowing simultaneous writes.

10   Fast Algorithms and Limits to Parallel Computation

In complexity theory a problem is said to be in the class P if it can be solved in sequential time $n^{O(1)}$, where n is the problem size. Similarly, a problem is said to be in the class NC if it can be solved in parallel time $(\log n)^{O(1)}$ using $n^{O(1)}$ processors, or in other words, if it can be solved in polylogarithmic time by using a polynomial number of processors.

Problems in P that are unlikely to be in NC are called *P-complete* problems and sometimes are also referred to as "inherently sequential", e.g., by Reif [61], who provided evidence that depth-first search is hard to parallelize. As we have already seen in Sect. 8, no inherently sequential problems exist, therefore referring to P-complete problems as such is inappropriate.

Problems in the class NC are the ones having fast parallel algorithms. Another important class of algorithms, called *cost-optimal*, or *work-optimal* algorithms, is the group algorithms with the product of the parallel running time and the number of processors used matching the sequential lower time bound for the problem. These algorithms with the parallel running time

$$t_p(n) = \frac{\Theta(t_1(n))}{p},$$

where n is the problem size, p is the number of processors and $t_1(n)$ is the running time of the best possible sequential algorithm for the problem, clearly break Amdahl's law.

There exist cost-optimal algorithms even for P-complete problems [13, 30, 69], e.g., Castanho et al. [13] devise cost-optimal algorithms for computational-geometry problems, such as the convex-layers and the envelope-layers problem.

On the other hand, Atallah et al. [7] prove that, unless $P = NC$, it is impossible to solve a number of two-dimensional geometric problems in polylogarithmic time by using a polynomial number of processors.

Greenlaw et al. [32] provide a thorough discussion of P-completeness theory and limits to parallel computation in general. However, these have nothing to do with Amdahl's law and its variants, which are not even mentioned in the book.

11 Discussion

Probably one of the most important contributions of this paper is the demonstration that sequential fractions of computations have negligible effect on speedup if the growth rate of the time requirement of the parallelizable fraction is higher than that of the sequential fraction. The successful application of GPUs for general-purpose computations [54] seems to justify that this is very often the case in practice.

Computations where the amount of sequential work grows with the number of processors, e.g., sequential concurrent objects [21] and mutual-exclusion algorithms [46], may appear in concurrent systems. However, these still have negligible effect on speedup if the growth rate of the parallelizable work is higher than the growth rate of the number of processors.

On the theoretical level, many researchers (including the author of this paper) published parallel algorithms ignoring sequential fractions, such as reading input, and therefore ignoring Amdahl's law. This concern was one of the motivations of this research. However, as long as the growth rate of the parallelizable fractions is higher, ignoring sequential fractions (including the ones needed, e.g., for synchronisation) is just the same as using the "big Oh" (Θ and Ω) notation for analysis of algorithms.

An unexpected result of this research is that, very likely, unnecessary attention has been paid to the assumed adverse effects of Amdahl's law, in order to justify asymmetric chip multiprocessor architectures, as summarised in Sect. 4. In addition to the disadvantages already mentioned there, asymmetric chip multiprocessors require more complicated system software. As Sun and Chen [65] properly conclude (though based on the wrong assumption of the validity of Gustafson's law) asymmetric multicore architectures would only be worth exploring if their symmetric counterparts could not deliver satisfactory performance.

Let us suppose that an^q and bn^r, respectively, are the time requirements of the sequential and the parallelizable fractions of an application, where a, b, q and r are positive constants and n is the problem size. Assuming a perfectly parallelizable fraction, the speedup achievable by p processors is

$$\frac{t_1(n)}{t_p(n)} = \frac{an^q + bn^r}{an^q + bn^r/p}.$$

If p approaches infinity, the upper limit for speedup is

$$\frac{t_1(n)}{t_p(n)} = 1 + \frac{b}{a}n^{r-q}. \tag{4}$$

For $r - q > 1$, we get polynomial speedup. If $r - q = 1$, we get the same as (3) in Sect. 8. If $r - q = 0$, we get Amdahl's law (since $\frac{a+b}{a} = \frac{1}{f}$).

However, as we have already seen, inherently sequential fractions do not exist in theory, and sequential fractions in practice are inherently different from parallelizable fractions, therefore $r-q = 0$ is hardly ever the case. Another reason why we should resist the temptation of proposing a generalisation of Amdahl's law is that (4) cannot take into account that in practice the time requirement of sequential fractions can be proportional to the number of processors.

While teaching concurrency, it would be tempting to rely on rules of thumb or simple scientific laws, such as (4). However, for proving upper and lower bounds [14, 24, 62] more sophisticated arguments are needed than the ones Amdahl's law and its variants are based on.

One of the most important open questions of computer science is to decide if $P = NC$. Amdahl's law, suggesting a constant upper limit for speedup, is more pessimistic than a "no" answer, while Gustafson's law, suggesting better speedup than achievable by highly parallelizabe problems, is more optimistic than a "yes" answer.

The results by Dymond and Tompa and Mak do not imply that $P = NC$, as Dymond et al. require exponential numbers of processors, but sufficient for the refutation of Amdahl's law, assuming an infinite number of processors. Therefore not only the question *"Should Amdahl's law be repealed?"* in the title of Preparata's 1995 invited talk [60] should be answered affirmatively, but also all the variants of Amdahl's law should be repealed.

12　Conclusions

We have demonstrated that ignoring the difference of the growth rates of the numbers of operations is one of the reasons why Amdahl's law and its variants fail in practice. We have also seen that inherently sequential computations, the cornerstone of Amdahl's law and its variants, do not exist in theory. Several authors observed that in practice parallel performance is limited by communication, synchronization and concurrent objects. We have seen that the time requirement of these can be proportional to the number of processors. However, we have demonstrated that such sequential fractions have negligible effect on speedup if the growth rate of the time requirement of the parallel fraction is higher than that of the sequential fraction.

We have also demonstrated that Gustafson's law, together with other variants of Amdahl's law, contradict established theoretical results. We can conclude that no simple formula or law governing concurrency exists. Experimental results should not be interpreted by rules of thumb, but theoretical results in mind.

Acknowledgements. The author thanks anonymous reviewers for their support and constructive criticism that helped to improve the presentation of the paper. One of the reviewers drew the author's attention to the fact that experimental results complement the author's demonstration that parallelizable fractions of computational loads may grow with the problem size. The author also appreciate comments by Shekhar Y. Borkar, John L. Hennessy and David A. Patterson.

References

1. Akl, S.G., Cosnard, M., Ferreira, A.G.: Data-movement-intensive problems: two folk theorems in parallel computation revisited. Theor. Comput. Sci. **95**(2), 323–337 (1992)
2. Alexandrov, A., et al.: MapReduce and PACT—comparing data parallel programming models. In: Härder, T., Lehner, W., Mitschang, B., Schöning, H., Schwarz, H. (eds.) Datenbanksysteme für Business, Technologie und Web (BTW), 14. Fachtagung des GI-Fachbereichs "Datenbanken und Informationssysteme" (DBIS), 2.-4.3.2011 in Kaiserslautern, Germany. LNI, vol. 180, pp. 25–44. GI (2011). http://subs.emis.de/LNI/Proceedings/Proceedings180/article10.html
3. Alikoski, H.A.: Über das Sylvestersche Vierpunktproblem. Suomalainen Tiedeakatemia (1938)
4. Amdahl, G.M.: Validity of the single processor approach to achieving large scale computing capabilities. In: Proceedings of Spring Joint Computer Conference, pp. 483–485. ACM, New York (1967)
5. Angel, E.: Interactive Computer Graphics: A Top-Down Approach Using OpenGL, 5th edn. Addison-Wesley Co., Inc., Pearson Education, Boston (2009)
6. Annavaram, M., Grochowski, E., Shen, J.: Mitigating Amdahl's law through EPI throttling. SIGARCH Comput. Archit. News **33**(2), 298–309 (2005). https://doi.org/10.1145/1080695.1069995
7. Atallah, M.J., Callahan, P.B., Goodrich, M.T.: P-complete geometric problems. Int. J. Comput. Geom. Appl. **03**(04), 443–462 (1993)
8. Borkar, S.: Thousand core chips: a technology perspective. In: Proceedings of the 44th Annual Design Automation Conference, DAC 2007, pp. 746–749. ACM, New York (2007). https://doi.org/10.1145/1278480.1278667
9. Borkar, S., Chien, A.A.: The future of microprocessors. Commun. ACM **54**, 67–77 (2011). https://doi.org/10.1145/1941487.1941507
10. Borkar, S.Y.: Personal communication (2017)
11. Boyd, C.: Data-parallel computing. Queue **6**(2), 30–39 (2008). https://doi.org/10.1145/1365490.1365499
12. Cai, G., Hu, W., Liu, G., Li, Q., Wang, X., Dong, W.: An effective speedup metric considering I/O constraint in large-scale parallel computer systems. In: 19th International Conference on Advanced Communication Technology (ICACT), pp. 816–822, February 2017
13. Castanho, C.D., Chen, W., Wada, K., Fujiwara, A.: Parallelizability of some P-complete geometric problems in the EREW-PRAM. In: Wang, J. (ed.) COCOON 2001. LNCS, vol. 2108, pp. 59–63. Springer, Heidelberg (2001). https://doi.org/10.1007/3-540-44679-6_7
14. Cook, S.A., Dwork, C.: Bounds on the time for parallel RAM's to compute simple functions. In: Proceedings of the 14th Annual ACM Symposium on Theory of Computing, STOC 1982, pp. 231–233. ACM, New York (1982)

15. Cook, S.A., Reckhow, R.A.: Time bounded random access machines. J. Comput. Syst. Sci. **7**(4), 354–375 (1973)
16. Dean, J., Ghemawat, S.: MapReduce: simplified data processing on large clusters. Commun. ACM **51**(1), 107–113 (2008)
17. Denning, P.J., Lewis, T.G.: Exponential laws of computing growth. Commun. ACM **60**(1), 54–65 (2017). https://doi.org/10.1145/2976758
18. Dévai, F.: An optimal hidden-surface algorithm and its parallelization. In: Murgante, B., Gervasi, O., Iglesias, A., Taniar, D., Apduhan, B.O. (eds.) ICCSA 2011. LNCS, vol. 6784, pp. 17–29. Springer, Heidelberg (2011). https://doi.org/10.1007/978-3-642-21931-3_2
19. Dévai, F.: Gustafson's law contradicts theory results (Letter to the Editor). Commun. ACM **60**(4), 8–9 (2017). https://doi.org/10.1145/3056859
20. Dymond, P.W., Tompa, M.: Speedups of deterministic machines by synchronous parallel machines. J. Comput. Syst. Sci. **30**(2), 149–161 (1985)
21. Ellen, F., Hendler, D., Shavit, N.: On the inherent sequentiality of concurrent objects. SIAM J. Comput. **41**(3), 519–536 (2012)
22. Esmaeilzadeh, H., Blem, E., Amant, R.S., Sankaralingam, K., Burger, D.: Power challenges may end the multicore era. Commun. ACM **56**(2), 93–102 (2013). https://doi.org/10.1145/2408776.2408797
23. Eyerman, S., Eeckhout, L.: Modeling critical sections in Amdahl's law and its implications for multicore design. SIGARCH Comput. Archit. News **38**(3), 362–370 (2010). https://doi.org/10.1145/1816038.1816011
24. Fich, F.E., Meyer auf der Heide, F., Ragde, P., Wigderson, A.: One, two, three ... infinity: lower bounds for parallel computation. In: Proceedings of the 17th Annual ACM Symposium on Theory of Computing, STOC 1985, pp. 48–58. ACM, New York (1985). https://doi.org/10.1145/22145.22151
25. Fich, F.E., Meyer auf der Heide, F., Wigderson, A.: Lower bounds for parallel random access machines with unbounded shared memory. Adv. Comput. Res. Parallel Distrib. Comput. **4**, 1–16 (1987)
26. Fich, F.E., Hendler, D., Shavit, N.: Linear lower bounds on real-world implementations of concurrent objects. In: Proceedings of the 46th Annual IEEE Symposium on Foundations of Computer Science, FOCS 2005, pp. 165–173. IEEE Computer Society, Washington, DC (2005). https://doi.org/10.1109/SFCS.2005.47
27. Forsell, M.: On the performance and cost of some PRAM models on CMP hardware. Int. J. Found. Comput. Sci. **21**(3), 387–404 (2010). https://doi.org/10.1142/S0129054110007325
28. Forsell, M.: A PRAM-NUMA model of computation for addressing low-TLP workloads. Int. J. Netw. Comput. **1**(1), 21–35 (2011). http://www.ijnc.org/index.php/ijnc/article/view/11
29. Fortune, S., Wyllie, J.: Parallelism in random access machines. In: Proceedings of the 10th Annual ACM Symposium on Theory of Computing, STOC 1978, pp. 114–118. ACM, New York (1978)
30. Fujiwara, A., Inoue, M., Masuzawa, T.: Parallelizability of some P-complete problems. In: Rolim, J. (ed.) IPDPS 2000. LNCS, vol. 1800, pp. 116–122. Springer, Heidelberg (2000). https://doi.org/10.1007/3-540-45591-4_14
31. Ghanim, F., Vishkin, U., Barua, R.: Easy PRAM-based high-performance parallel programming with ICE. IEEE Trans. Parallel Distrib. Syst. **29**, 377–390 (2018)
32. Greenlaw, R., Hoover, H.J., Ruzzo, W.L.: Limits to Parallel Computation: P-Completeness Theory. Oxford University Press Inc., New York (1995)
33. Gustafson, J.L.: Reevaluating Amdahl's law. Commun. ACM **31**(5), 532–533 (1988)

34. Hennessy, J.L., Patterson, D.A.: Computer Architecture: A Quantitative Approach, 5th edn. Morgan Kaufmann Publishers Inc., San Francisco (2011)
35. Hennessy, J.L.: Personal communication (2017)
36. Herlihy, M., Shavit, N.: The Art of Multiprocessor Programming, Revised Reprint, 1st edn. Morgan Kaufmann Publishers Inc., San Francisco (2012)
37. Hill, M.D., Marty, M.R.: Amdahl's law in the multicore era. Computer 41(7), 33–38 (2008)
38. Hillis, W.D., Steele Jr., G.L.: Data parallel algorithms. Commun. ACM 29(12), 1170–1183 (1986)
39. Juurlink, B., Meenderinck, C.H.: Amdahl's law for predicting the future of multi-cores considered harmful. SIGARCH Comput. Archit. News 40(2), 1–9 (2012)
40. Karloff, H., Suri, S., Vassilvitskii, S.: A model of computation for MapReduce. In: Proeedings of the 21st Annual ACM-SIAM Symposium on Discrete Algorithms, SODA 2010, pp. 938–948. Society for Industrial and Applied Mathematics, Philadelphia (2010). http://dl.acm.org/citation.cfm?id=1873601.1873677
41. Karp, A.H., Flatt, H.P.: Measuring parallel processor performance. Commun. ACM 33(5), 539–543 (1990). https://doi.org/10.1145/78607.78614
42. Karp, R.M., Ramachandran, V.: Parallel algorithms for shared-memory machines. In: Handbook of Theoretical Computer Science, vol. A, pp. 869–941. MIT Press, Cambridge (1990). http://portal.acm.org/citation.cfm?id=114872.114889
43. Kucera, L.: Parallel computation and conflicts in memory access. Inf. Process. Lett. 14(2), 93–96 (1982). https://doi.org/10.1016/0020-0190(82)90093-X
44. Kuck, D.J.: A survey of parallel machine organization and programming. ACM Comput. Surv. 9(1), 29–59 (1977). https://doi.org/10.1145/356683.356686
45. Kumar, R., Tullsen, D.M., Jouppi, N.P., Ranganathan, P.: Heterogeneous chip multiprocessors. Computer 38(11), 32–38 (2005)
46. Lamport, L.: A new solution of Dijkstra's concurrent programming problem. Commun. ACM 17(8), 453–455 (1974). https://doi.org/10.1145/361082.361093
47. Luccio, F., Pagli, L.: The p-shovelers problem: (computing with time-varying data). SIGACT News 23, 72–75 (1992)
48. Luebke, D., Humphreys, G.: How GPUs work. Computer 40, 96–100 (2007). http://dl.acm.org/citation.cfm?id=1251557.1251701
49. Mak, L.: Parallelism always helps. SIAM J. Comput. 26(1), 153–172 (1997)
50. Marowka, A.: Energy-aware modeling of scaled heterogeneous systems. Int. J. Parallel Program. 45, 1–20 (2017)
51. McKenna, M.: Worst-case optimal hidden-surface removal. ACM Trans. Graph. 6, 19–28 (1987)
52. Mittal, S.: A survey of techniques for architecting and managing asymmetric multicore processors. ACM Comput. Surv. 48(3), 45:1–45:38 (2016). https://doi.org/10.1145/2856125
53. Morad, A., Yavits, L., Kvatinsky, S., Ginosar, R.: Resistive GP-SIMD processing-in-memory. ACM Trans. Archit. Code Optim. 12(4), 57:1–57:22 (2016). https://doi.org/10.1145/2845084
54. Nickolls, J., Buck, I., Garland, M., Skadron, K.: Scalable parallel programming with CUDA. Queue 6(2), 40–53 (2008). https://doi.org/10.1145/1365490.1365500
55. Patterson, D., Hennessy, J.: Computer Organization and Design: The Hardware/Software Interface. The Morgan Kaufmann Series in Computer Architecture and Design. Elsevier Science, $ARM^{\circledR}$ edn. (2016)
56. Patterson, D.A.: Personal communication (2017)
57. Patterson, D.A., Gibson, G., Katz, R.H.: A case for redundant arrays of inexpensive disks (RAID). SIGMOD Rec. 17(3), 109–116 (1988)

58. Paul, J.M., Meyer, B.H.: Amdahl's law revisited for single chip systems. Int. J. Parallel Program. **35**(2), 101–123 (2007). https://doi.org/10.1007/s10766-006-0028-8
59. Philip, J.: The area of a random triangle in a square. Technical report TRITA MAT 10 MA 01, Royal Institute of Technology (2010). http://www.math.kth.se/~johanph/squaref.pdf
60. Preparata, F.P.: Should Amdahl's Law be repealed? In: Staples, J., Eades, P., Katoh, N., Moffat, A. (eds.) ISAAC 1995. LNCS, vol. 1004, pp. 311–311. Springer, Heidelberg (1995). https://doi.org/10.1007/BFb0015436
61. Reif, J.H.: Depth-first search is inherently sequential. Inf. Process. Lett. **20**(5), 229–234 (1985)
62. Roughgarden, T., Vassilvitskii, S., Wang, J.R.: Shuffles and circuits: (on lower bounds for modern parallel computation). In: Proceedings of the 28th ACM Symposium on Parallelism in Algorithms and Architectures, SPAA 2016, pp. 1–12. ACM, New York (2016). https://doi.org/10.1145/2935764.2935799
63. Shavit, N.: Data structures in the multicore age. Commun. ACM **54**, 76–84 (2011). https://doi.org/10.1145/1897852.1897873
64. Suleman, M.A., Mutlu, O., Qureshi, M.K., Patt, Y.N.: Accelerating critical section execution with asymmetric multi-core architectures. SIGPLAN Not. **44**(3), 253–264 (2009). https://doi.org/10.1145/1508284.1508274
65. Sun, X.H., Chen, Y.: Reevaluating Amdahl's law in the multicore era. J. Parallel Distrib. Comput. **70**(2), 183–188 (2010)
66. Valiant, L.G.: Parallelism in comparison problems. SIAM J. Comput. **4**(3), 348–355 (1975). https://doi.org/10.1137/0204030
67. Valiant, L.G.: A bridging model for parallel computation. Commun. ACM **33**(8), 103–111 (1990). https://doi.org/10.1145/79173.79181
68. Vishkin, U.: A PRAM-on-chip vision (invited abstract). In: Proceedings of the Seventh International Symposium on String Processing Information Retrieval (SPIRE 2000), p. 260. IEEE Computer Society, Washington, DC (2000). http://portal.acm.org/citation.cfm?id=829519.830820
69. Vitter, J.S., Simons, R.A.: New classes for parallel complexity: a study of unification and other complete problems for P. IEEE Trans. Comput. **35**(5), 403–418 (1986). https://doi.org/10.1109/TC.1986.1676783
70. White, T.: Hadoop: The Definitive Guide. O'Reilly Media Inc., Sebastopol (2012)
71. Woo, D.H., Lee, H.H.: Extending Amdahl's law for energy-efficient computing in the many-core era. Computer **41**(12), 24–31 (2008)
72. Yavits, L., Morad, A., Ginosar, R.: The effect of communication and synchronization on Amdahl's law in multicore systems. Parallel Comput. **40**(1), 1–16 (2014). https://doi.org/10.1016/j.parco.2013.11.001
73. Yavits, L., Morad, A., Ginosar, R.: The effect of temperature on Amdahl law in 3D multicore era. IEEE Trans. Comput. **65**(6), 2010–2013 (2016). https://doi.org/10.1109/TC.2015.2458865

A Distance Matrix Completion Approach to 1-Round Algorithms for Point Placement in the Plane

Md. Zamilur Rahman[1], Udayamoorthy Navaneetha Krishnan[1], Cory Jeane[1], Asish Mukhopadhyay[1(✉)], and Yash P. Aneja[2]

[1] School of Computer Science, University of Windsor,
Windsor, ON N9B 3P4, Canada
`asishm@uwindsor.ca`
[2] Odette School of Business, University of Windsor, Windsor, ON N9B 3P4, Canada

Abstract. In this paper we propose a 1-round algorithm for approximate point placement in the plane in an adversarial model. The distance query graph presented to the adversary is chordal. The remaining distances are determined using a distance matrix completion algorithm for chordal graphs, based on a result by Bakonyi and Johnson [SIAM Journal on Matrix Analysis and Applications 16 (1995)]. The layout of the points is determined from the complete distance matrix in two ways: using the traditional Young-Householder approach as well as a Stochastic Proximity Embedding (SPE) method due to Agrafiotis [Journal of Computational Chemistry 24 (2003)].

Keywords: Distance geometry · Point placement
Distance matrix completion · Embed algorithm
Eigenvalue decomposition

1 Introduction

The problem of locating n distinct points on a line, up to translation and reflection, in an adversarial setting has been extensively studied [1–4]. The best known 2-round algorithm that makes $9n/7$ queries and has a query lower bound of $9n/8$ queries is due to Alam and Mukhopadhyay [5]. In this paper we propose a 1-round algorithm for the same problem in the plane. To the best of our knowledge there is no prior work extant on this problem. A practical motivation for this study is the extensively researched and closely related sensor network localization problem [6,7].

2 Preliminaries

Let $D = [d_{ij}]$ be an $n \times n$ symmetric matrix, whose diagonal entries are 0 and the off-diagonal entries are positive. It is said to be an Euclidean distance

© Springer-Verlag GmbH Germany, part of Springer Nature 2018
M. L. Gavrilova and C. J. K. Tan (Eds.): Trans. on Comput. Sci. XXXIII,
LNCS 10990, pp. 97–114, 2018.
https://doi.org/10.1007/978-3-662-58039-4_6

matrix if there exists points $p_1, p_2, \ldots, p_n$ in some k-dimensional Euclidean space such that $d_{ij} = d(p_i, p_j)^2$, where $d(p_i, p_j)$ is the Euclidean distance between the points p_i and p_j. A set of necessary and sufficient conditions for this was given by Schoenberg [8], as well as Young and Householder [9]. A partial distance matrix is one in which some entries are missing.

A *graph* G consists of a finite set of vertices (also called nodes) $\{v_1, v_2, \ldots, v_n\}$ and a set of edges $\{\{v_i, v_j\}, i \neq j\}$ joining some pairs of vertices. A standard description is $G = (V, E)$, where V is the set of vertices set and E is the set of edges. A *path* in G is a sequence of vertices $v_i, v_{i+1}, \ldots, v_k$, where $\{v_j, v_{j+1}\}$ for $j = i, i+1, \ldots, k-1$, is an edge of G. A *cycle* is a closed path. The *size* of a cycle is the number of edges in it. A *chord* of a cycle is an edge joining two non-consecutive vertices. A graph G is said to be *chordal* if it has no chordless cycles of size 4 or more.

The *distance graph* of an $n \times n$ distance matrix, is a graph on n vertices with an edge connecting two vertices v_i and v_j if there is a non-zero entry in i-th row and j-th column of the distance matrix.

The *neighbourhood* $N(v)$ of a vertex v of G consists of those vertices in G that are adjacent to v. A vertex v is said to be *simplicial* if $N(v)$ is a clique, that is, a complete subgraph on $N(v)$. A *simplicial ordering* of the vertices of G is a map $\alpha : V \to \{1, 2, \ldots, n\}$ such that v_i is simplicial in the induced graph on the the vertex set $\{v_i, v_{i+1}, \ldots, v_n\}$.

3   Point Placement on a Line: A Quick Review

To provide a context and motivation for the results of this paper, we provide a quick review of the main ideas underlying point placement algorithms for points on a line, with reference to a state-of-the-art algorithm [5].

Let $P = \{p_1, p_2, \ldots, p_n\}$ be n distinct points on a line. A distance graph on n vertices (corresponding to the n points in P) has edges joining pairs of points whose distances on the line are sought of an adversary. An assignment of lengths to the edges of this graph by an adversary is assumed to be valid if there exists a linear layout consistent with these lengths. The distance graph is said to be *line-rigid* if a consistent layout exists for all valid, adversarial assignments of lengths. All the distance graphs shown in Fig. 1 are line-rigid. However, a 4-cycle is not line-rigid as there exists an assignment of lengths that makes it a parallelogram, whose vertices have two distinct linear layouts.

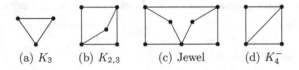

(a) K_3 (b) $K_{2,3}$ (c) Jewel   (d) K_4^-

Fig. 1. Some examples of line rigid graphs.

A prototypical 1-round algorithm uses the line-rigid 3-cycle (or triangle) graph as the core structure and constructs the following distance graph on n points (see Fig. 2). As the figure shows, the graph has $n - 2$ triangles, hanging from a common strut. The number of distance queries made is $2n - 3$.

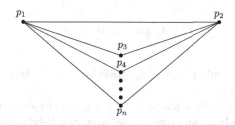

Fig. 2. Distance graph for a 1-round algorithm

A prototypical 2-round algorithm uses the 4-cycle (or triangle) graph as the core structure and constructs the following distance graph on n points (see Fig. 3). As the figure shows, the graph has b and $b + 2$ edges, hanging from the left and right end-points respectively of a fixed edge. The explanation is that a 4-cycle is not line-rigid and the rigidity condition that at least one pair of opposite edges are not equal can be satisfied over two rounds. The discrepancy of 2 in the number of edges hanging from the two end-points allows us to pair edges which are not equal and thus meet the line rigidity condition for a 4-cycle. The number of distance queries made is $3n/2 - 2$. Thus by increasing the number of rounds and constructing a more complex query graph we reduce the number of distance queries by a constant factor.

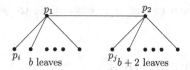

Fig. 3. Distance graph for a 2-round algorithm

The best known 2-round algorithm to-date [5], builds a distance query graph using the 3-path graph of Fig. 4. Its query complexity is $9n/7 + O(1)$. This comes at the expense of 55 rigidity conditions that must be satisfied over two rounds.

The main tool for obtaining these rigidity conditions is the concept of a layer graph, introduced in [3]. A layer graph is an orthogonal re-drawing (if possible) of the distance query graph that must satisfy the following conditions:

P1. Each edge e of G is parallel to one of the two orthogonal directions $\mathbf{x}$ and $\mathbf{y}$.
P2. The length of an edge e is the distance between the corresponding points on L.

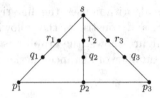

Fig. 4. The 3-path graph

P3. Not all edges are along the same direction (thus a layer graph has a two-dimensional extent).
P4. When the layer graph is folded onto a line, by a rotation either to the left or to the right about an edge of the layer graph lying on this line, no two vertices coincide.

Chin et al. [3] showed that a given distance query graph is not line rigid iff it has a layer graph drawing. The different rigidity conditions are derived from a painstaking enumeration of all possible layer graph drawings of the given distance query graph. This is a challenging task.

Our experience with the implementation of 2-round algorithms (see [10]) has shown that the rigidity conditions are easy to verify when exact arithmetic is used; indeed, we simulated an adversary by generating layouts with integral coordinates. However if pairwise distances are not integral, the rounding errors introduced in finite-precision calculations can make checking the rigidity conditions difficult. This is an unavoidable issue for point-placement in the plane.

Another difficulty of generalizing this approach to two and higher dimensions is that of obtaining a suitable generalization of the layer graph concept and the associated theorem. This motivates the approach taken in this paper. The advantage of this approach is that it is susceptible to generalization to higher dimensions.

4   Overview of Our Results

We first discuss a reductionist approach to this problem: reducing point placement in the plane to point placement on a line. We consider the case when the points lie on a circle, using stereographic projection to reduce this to a 1-dimensional point placement problem. For points lying on an integer grid, we reduce the problem to two 1-dimensional point placement problems.

The algorithms for point placement on a line require testing a very large number of constraints involving edge lengths of a distance graph. Our experiments have shown that these work well when the points on a line have integral coordinates. To circumvent this problem we consider a matrix distance completion approach, when the distance graph is chordal. In our adversarial setting, we seek the lengths of this chordal graph from an adversary (An adversary can be thought of as a source of correct distance measurements.).

Once the adversary has returned edge lengths for the chordal distance graph, we solve a matrix distance completion problem. Bakonyi and Johnson [11] showed that if the distance graph corresponding to a partial distance matrix is chordal, there exists a completion of this partial distance matrix. Precisely, they proved the following result.

Theorem 1. [11] *Every partial distance matrix in R^k, the graph, G, of whose specified entries is chordal, admits a completion to a distance matrix in R^k.*

Finally, we compute the planar coordinates of the vertices of this complete distance graph. We have used two methods for this: an intriguing heuristic known as Stochastic Proximity Embedding (SPE) [12] and a deterministic technique due to Young and Householder [9].

5   Point Placement in the Plane

When the points $p_1, p_2, \ldots, p_n$ lie on an integer grid, we can solve the problem by solving two 1-dimensional point placement problems by projecting them on the x and y-axes. We assume that no two points lie on the same vertical or horizontal line of the grid (see Fig. 5).

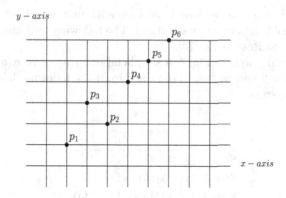

Fig. 5. Points on a two-dimensional integer grid

When the points lie on a circle, we can solve the problem by stereographic projection of the points on a line and then applying a 1-dimensional point location algorithm to the projected points (see Fig. 6).

When the (distance) graph of the partial distance matrix is chordal, we use a distance matrix completion algorithm, the major components of which are discussed below.

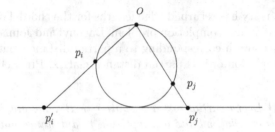

Fig. 6. Stereographic projection of points on a circle

Algorithm 1. Simplicial Ordering

1: Assign the empty label list, (), to each vertex in V
2: **for** $i = n$ downto 1 **do**
3:     Pick a vertex $v \in V$ with the lexicographically largest label list
4:     Set $\alpha(v) = i$
5:     For each unnumbered vertex w adjacent to v, add i to the label list of w
6: **end for**
7: return α

5.1   Computing a Simplicial Ordering, α, of G

A simplicial ordering can be found by a breadth-first search of G, combined with a lexicographic labeling of its vertices. The following well-known LEX-BFS algorithm is due to Rose et al. [13].

For the chordal graph of Fig. 7, a simplicial ordering of is: $\alpha(a) = 5, \alpha(b) = 4, \alpha(c) = 3, \alpha(e) = 2$ and $\alpha(d) = 1$, obtained from the following label lists of the vertices over five steps.

	a	b	c	d	e
Step 0	()	()	()	()	()
Step 1	()	(5)	(5)	()	()
Step 2	()	(5)	(5,4)	(4)	(4)
Step 3	()	(5)	(5,4)	(4)	(4,3)
Step 4	()	(5)	(5,4)	(4,2)	(4,3)

5.2   Computing a Chordal Graph Sequence

An algorithm for generating the sequence of chordal graphs depends on the following results, proved in [14].

Theorem 2. [14] *G has no minimal cycles of length exactly 4 if and only if the following holds: For any pair of vertices u and v with $u \neq v$, $\{u, v\} \notin E$, the*

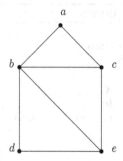

Fig. 7. A chordal graph on five vertices

graph $G + \{u, v\}$ *has a unique maximal clique which contains both u and v. (That is: if C and C' are both cliques in $G + \{u, v\}$ which contain u and v, then so is $C \cup C'$.)*

In particular, Theorem 2 holds for chordal graphs. The next theorem suggests an iterative algorithm for solving the distance matrix completion problem.

Theorem 3. [14] *Let $G = (V, E)$ be chordal. Then there exists a sequence of chordal graphs $G_i = (V, E_i)$, $i = 0, 1, \ldots, k$, such that $G = G_0, G_1, G_2, \ldots, G_k$ is the complete graph and G_i is obtained by adding to G_{i-1} an edge $\{u, v\}$ as in Theorem 2.*

Such an edge $\{u, v\}$ is selected using the following scheme described in [14]. Assume that a simplicial ordering α of the vertices of the input chordal graph G is available. Let v_k be the vertex $\alpha^{-1}(k)$. Set $k_i = max\{k | (v_k, v_m) \notin E_i$ for some $m\}$ and $r_i = max\{r | (v_r, v_{k_i}) \notin E_i\}$. Then the edge to be added is $\{u, v\} = \{u_{k_i}, v_{r_i}\}$. In the next section we discuss an algorithm for selecting a maximal clique, containing this edge.

5.3   Computing a Maximal Clique Containing a Given Edge

An interesting algorithm due to Bron-Kerbosch [15] computes all maximal cliques, from which we can select the maximal clique that contains this edge. The Bron-Kerbosch algorithm is a recursive backtracking algorithm, and a version based on choosing a pivot is described thus. The algorithm maintains three sets R, P and X, reporting the set R as the vertices of a maximum clique when at any level of the recursive calls, the sets P and X become empty.

We have implemented a simple algorithm that starts with the edge of interest and grows this into a maximal clique. In greater details, start with a clique containing two vertices of the given edge, and grow the current clique one vertex at a time by looping through the graph's remaining vertices. For each vertex v examined, add v to the clique if it is adjacent to every vertex that is already in the clique; otherwise, discard v.

Algorithm 2. Computing Cliques

1: BronKerbosch2(R, P, X):
2: **if** (P and X are both empty) **then**
3:     report R as a maximal clique
4: **end if**
5: choose a pivot vertex $u \in P \cup X$
6: **for** each vertex $v \in P \setminus N(u)$ **do**
7:     BronKerbosch2$(R \cup \{v\}, P \cap N(v), X \cap N(v))$
8:     $P := P \setminus \{v\}$
9:     $X := X \cup \{v\}$
10: **end for**

5.4   Distance Matrix Completion of a Clique

The distance matrix of a clique with the distance of one edge missing can be formulated as the problem of completing a partial distance matrix with one missing entry. The following lemma proposes a solution to this problem.

Theorem 4. [11] *The partial distance matrix*

$$\begin{pmatrix} 0 & D_{12} & x \\ D_{12}^t & D_{22} & D_{23} \\ x & D_{23}^t & 0 \end{pmatrix}$$

admits at least one completion to a distance matrix F. Moreover, if

$$\begin{pmatrix} 0 & D_{12} \\ D_{12}^t & D_{22} \end{pmatrix}$$

and

$$\begin{pmatrix} D_{22} & D_{23} \\ D_{23}^t & 0 \end{pmatrix}$$

are distance matrices with embedding dimensions p and q then x can be chosen so that the embedding dimension of F is $s = max\{p, q\}$.

This is equivalent to finding completions of the partial distance matrix:

$$\begin{pmatrix} 0 & 1 & 1 & e^t & 1 \\ 1 & 0 & d_{12} & \overline{D}_{13} & d_{14} \\ 1 & d_{12} & 0 & \overline{D}_{23} & x \\ e \, \overline{D}_{13}^t & \overline{D}_{23}^t & \overline{D}_{33} & \overline{D}_{34} \\ 1 & d_{14} & x & \overline{D}_{34}^t & 0 \end{pmatrix}$$

to a matrix in which the Schur complement

$$\begin{pmatrix} a & B & x - d_{12} - d_{14} \\ B^t & C & D \\ x - d_{12} - d_{14} & D^t & f \end{pmatrix}$$

of the upper left 2×2 principal submatrix

$$\begin{pmatrix} 0 & 1 \\ 1 & 0 \end{pmatrix}$$

has a positive semidefinite completion of rank s. This provides a solution for x that follows from the following result.

Theorem 5. [16] *Let*

$$R = \begin{pmatrix} a & B & x \\ B^t & C & D \\ x & D^t & f \end{pmatrix}$$

be a real partial positive semidefinite matrix in which $rank \begin{pmatrix} a & B \\ B^t & C \end{pmatrix} = p$ *and* $rank \begin{pmatrix} C & D \\ D^t & f \end{pmatrix} = q$. *Then there is real positive semidefinite completion F of R such that the rank of F is $max\{p, q\}$. This completion is unique iff $rankC = p$ or $rankC = q$.*

In the next section we discuss the last stage of the point placement problem: this is to determine the coordinates of a layout from a completed distance matrix.

6 Coordinate Computation

We apply and compare two methods: an intriguing heuristic known as Stochastic Proximity Embedding (SPE) [12] and a deterministic technique due to Young and Householder [9]. Below we describe these briefly.

6.1 The SPE Method

Let $P^r = \{p_1^a, p_2^a, \ldots, p_n^a\}$ be an arbitrary embedding in two-dimensions of the points $P = \{p_1, p_2, \ldots, p_n\}$. The input to the algorithm is a complete (or incomplete) distance matrix $R = [r_{ij}]$, representing the exact distances between some pairs of points as well as an approximate distance matrix $D = [d_{ij}]$, representing approximate distances between all pairs of points. The latter matrix is derived from the arbitrary embedding $\{p_1^a, p_2^a, \ldots, p_n^a\}$.

The algorithm is surprisingly simple, consisting of two nested cycles: an outer cycle, called the learning cycle and an inner cycle that applies a Newton-Raphson root-finding style correction to the coordinates of a randomly chosen pair of points from the set P^r. The number of iteration steps is $C \times S$, where C is the number of learning cycles and S is the number of times a random pair of points is selected from P^r. The parameters C and S are set so that $CS = o(n^2)$, as the quadratic running times of similar algorithms based on Multi-Dimensional Scaling (MDS), Principal Component Analysis (PCA), for example, make them

Algorithm 3. SPE

[12]

1: Initialize the coordinates of P^r
2: Select a pair of points, p_i^a and p_j^a, at random and compute their distance $d_{ij} = ||p_i^a - p_j^a||$.
3: **if** $(d_{ij} \neq r_{ij})$ **then**
4:     $p_i^a \leftarrow p_i^a + \lambda \frac{1}{2} \frac{r_{ij} - d_{ij}}{d_{ij} + \epsilon} (p_i^a - p_j^a), \epsilon \neq 0$
5:     $p_j^a \leftarrow p_j^a + \lambda \frac{1}{2} \frac{r_{ij} - d_{ij}}{d_{ij} + \epsilon} (p_j^a - p_i^a), \epsilon \neq 0$
6: **end if**
7: Repeat Step 2 for a prescribed number of steps, S
8: Decrease the learning rate λ by prescribed decrement $\delta\lambda$
9: Repeat Steps 2-4 for a prescribed number of cycles, C

inefficient when dealing with very large point sets. Algorithm 3 SPE is a formal description of the SPE heuristic.

During our experiments with the SPE heuristic, we observed that if the number of points to be learned is fewer than the number of learning cycles, SPE ends up translating the point set by a very large value. To preempt this translation, we introduced the notion of anchor points. These are a few points whose positions are never updated, treating them, so to say, as fixed points of the transformation. We modified the above original version of SPE accordingly.

6.2   The Young-Householder Method

Young and Householder's algorithm [9] for coordinate computation is based on the following result.

Theorem 6. [9] *A necessary and sufficient condition for a set of numbers $d_{ij} = d_{ji}$ to be the mutual distances of a real set of points in Euclidean space is that the matrix $B = [d_{1i}^2 + d_{1j}^2 - d_{ij}^2]$ be positive semi-definite; and in this case the set of points is unique apart from a Euclidean transformation.*

In this case, there exists an orthogonal matrix σ such that

$$B = \sigma L^2 \sigma^t$$

where $L^2 = [\lambda_1^2, \lambda_2^2, \ldots, \lambda_r^2, 0, \ldots, 0]$, σ^t is the transpose of σ, and r is the embedding dimension. Thus, we have

$$B = (\sigma L)(\sigma L)^t$$

Since $B = AA^t$, where the rows of the matrix A are the coordinates of the points $p_1, p_2, \ldots, p_n$ in some r dimensional Euclidean space, the coordinates of the points are determined by solving the system of linear equations.

$$A = \sigma L$$

7   SPE Versus Young-Householder

For the SPE heuristic, we set the so-called R-matrix to the completed distance matrix. To estimate the match of the layouts, we fixed one of the layouts and performed the geometric transformations (translations and rotations) on the other layouts and then we computed their Root-mean-square deviation (RMSD) distances. We computed RMSD distances using Eq. 1.

$$RMSD(p,q) = \sqrt{\frac{1}{n} \sum_{i=1}^{n} ((p_{ix} - q_{ix})^2 + (p_{iy} - q_{iy})^2)} \tag{1}$$

Here we applied Kabsch algorithm [17] to perform a translation and the optimal rotation. For translation, we computed the centroids from both sets of coordinates and then subtracted the centroids from all the points in each set. Then we computed the covariance matrix. Next we computed singular value decomposition (SVD) of the covariance matrix and after we computed the optimal rotation matrix.

Remarks: During our experiments with the SPE heuristic, we observed that if the number of points to be learned is fewer than the number of learning cycles, SPE ends up translating the point set by a very large value. This translation can be preempted by assuming that the positions of a few points, named as anchor points, are known a priori. Such an assumption is not very far-fetched as in the network localization problem the positions of some sensors, called beacon nodes are known. Inside the loops of the SPE heuristic, the positions of the anchor points are never updated, treating them, so to say, as fixed points of the transformation. Anchor points have another merit: SPE becomes less sensitive to the initial random distribution we start with. Otherwise, SPE has to be run with different initial distributions to get a final layout that agrees closely with the layout generated by the Young-Householder method.

7.1   Experimental Results

We have implemented the above algorithm in Python on a laptop with an Intel Core i7 processor with 16GB RAM, running under Windows 10. The software includes a module for the generation of chordal graphs to be used as input. The chordal graph generation uses one of two algorithms proposed by Markenzon et al. [18]. A formal description of this algorithm is given below.

The result of an experiment is described below for a chordal graph. The following partial distance matrix, where the off-diagonal 0's represent unknown

---

Algorithm 4. ChordalGraphGeneration(G)

1: Generate a binary tree, G
2: Pick two random vertices u and v from G.
3: **if** there is an edge exists between these two vertices (u and v) **then**
4:     go back to step 2.
5: **else**
6:     find the neighbors of u and v.
7: **end if**
8: **if** there is no common neighbor between u and v **then**
9:     go back to step 2.
10: **else**
11:     choose one vertex x from the common neighbor of u and v
12: **end if**
13: Perform BFS from u to v on the graph consisting of all the adjacent vertices of x minus the intersection of u and v's neighbors
14: **if** BFS finds a path from u to v **then**
15:     go back to step 2.
16: **else**
17:     insert an edge between u and v
18: **end if**

---

distances,

$$\begin{pmatrix} 0 & 9 & 0 & 0 & 0 & 0 & 5 & 20 \\ 9 & 0 & 2 & 25 & 40 & 34 & 20 & 17 \\ 0 & 2 & 0 & 17 & 0 & 0 & 0 & 13 \\ 0 & 25 & 17 & 0 & 5 & 0 & 0 & 2 \\ 0 & 40 & 0 & 5 & 0 & 2 & 0 & 5 \\ 0 & 34 & 0 & 0 & 2 & 0 & 10 & 5 \\ 5 & 20 & 0 & 0 & 0 & 10 & 0 & 13 \\ 20 & 17 & 13 & 2 & 5 & 5 & 13 & 0 \end{pmatrix}$$

is obtained from the distances returned by the adversary based on the layout of a chordal graph, G, shown in Fig. 8.

The matrix distance completion algorithm outputs the following matrix.

$$\begin{pmatrix} 0 & 9 & 17.0 & 34.0 & 37.0 & 25.0 & 5 & 20 \\ 9 & 0 & 2 & 25 & 40 & 34 & 20 & 17 \\ 17.0 & 2 & 0 & 17 & 34.0 & 32.0 & 26.0 & 13 \\ 34.0 & 25 & 17 & 0 & 5 & 9.0 & 25.0 & 2 \\ 37.0 & 40 & 34.0 & 5 & 0 & 2 & 20.0 & 5 \\ 25.0 & 34 & 32.0 & 9.0 & 2 & 0 & 10 & 5 \\ 5 & 20 & 26.0 & 25.0 & 20.0 & 10 & 0 & 13 \\ 20 & 17 & 13 & 2 & 5 & 5 & 13 & 0 \end{pmatrix}$$

where the computed entries are shown with a decimal point, followed by a single 0. The correctness of the computed entries can be checked against Fig. 8.

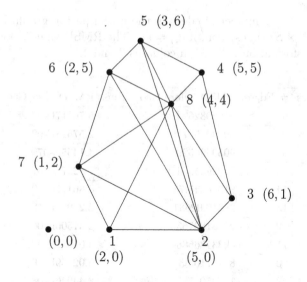

Fig. 8. A chordal graph on 8 vertices

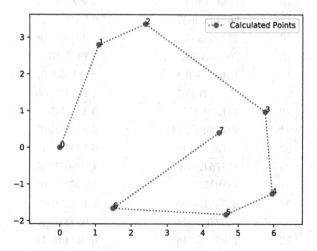

Fig. 9. A plot of the output of the Young-Householder algorithm

We ran the above complete distance matrix through our implementation of the Young-Householder algorithm. A plot of the output is shown in Fig. 9. Next we ran the SPE algorithm and the embedding is shown in Fig. 10. The embedding after translation and rotation is shown in Fig. 11. We plot the YH vs SPE in Fig. 12 where we can see the embedding from both methods are same. The RMSD value before and after geometric transformation is 4318.60872515 and $3.11126009039e^{-13}$, respectively.

We ran experiments on different chordal graphs and listed the results in Sect. 7.1. In this table, #Vertices refers to the number of vertices in the graph,

#Edges refers to the number of edges in the initial partial graph. For a graph, we ran SPE $(C \times S)$ times or until $d_{ij} = r_{ij}$. The RMSD value before and after geometric transformation is listed in columns 3 and 4.

# Vertices	# Edges	RMSD before Geo. Trans	RMSD after Geo. Trans
4	5	6164982694	5.76717E-07
5	9	304301527.4	2.57528E-08
5	8	9059590529	4.1455E-07
5	7	1222955924	7.82276E-07
6	14	81556345.73	8.86177E-09
6	13	28010.02681	2.23914E-12
6	12	29981807636	2.31506E-06
6	11	48331865860	3.01052E-06
7	20	8577710.63	5.02859E-10
7	19	1273505726	8.44946E-08
7	18	8261513997	8.99897E-07
7	17	663158.4682	6.56741E-11
7	16	318999605.8	3.63369E-08
7	15	2006376531	1.745E-07
7	14	121336040.3	7.96895E-07
8	27	177582109.7	1.17449E-08
8	26	551350.4771	3.46512E-11
8	24	4.67258E+11	3.97418E-05
8	23	907371.8146	6.12932E-11
9	35	8737642714	1.03805E-06
10	44	188935050.7	1.73626E-08
11	54	88538220.03	8.17251E-09
11	53	844370.2674	1.05934E-10
11	52	1196485.246	6.64448E-11
12	65	31394552.32	3.60513E-09
12	64	6083483.794	3.78314E-30
12	63	463247953.3	2.1168E-08
13	77	24104264.98	6.6898E-08
13	76	97530293	1.59546E-08
13	75	686054.2491	8.1589E-08
13	74	18470126.97	1.90852E-09
13	73	2062075.845	1.40785E-07
14	90	2847630.879	2.26219E-10

# Vertices	# Edges	RMSD before Geo. Trans	RMSD after Geo. Trans
15	104	2117864.588	2.65562E-10
15	103	3658041.023	2.5549E-08
16	118	40338688.31	6.72352E-07
17	135	2300705.974	8.44445E-10
17	134	1171271.563	2.64047E-10
17	133	2726784111	3.98553E-07
18	152	101220497.1	2.63716E-08
19	170	23454858.87	9.07134E-05
19	169	2297515.954	3.50958E-09
19	168	186898258.3	1.71402E-08
20	189	5015786.442	2.16286E-08
20	188	710440.9778	1.85761E-05
20	187	222145709.8	2.3955E-07
20	186	22051830.05	0.000203226
20	185	284561.9915	3.98301E-11
20	184	10035500.9	0.149877199
21	209	61660621.28	5.02581E-09
21	208	1988789.434	1.60774E-07
22	230	14488854.69	0.000815554
22	228	11976883.32	0.001114898
22	227	8297614.582	0.438067978
23	251	430611.4293	1.50913E-10
24	275	3.99943E+11	0.05391485
25	299	16517667.05	0.353860889
25	298	118930088.8	0.040632086
26	323	8397700.33	1.28898E-09
27	350	139031643.7	2.61083E-06
28	377	425730.4965	0.000547247
30	434	18810003.15	1.000356519
30	433	2266368.665	0.011282119
30	432	1187155.942	1.67458E-07
30	430	41982908.09	6.05155E-07
31	463	963614.5917	1.82794E-08
32	494	378732.6461	7.2889E-05
32	493	13433060.45	2.43305E-07
33	525	2811847.373	2.01482E-09

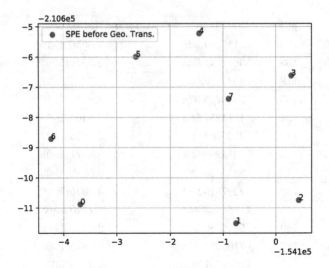

Fig. 10. A plot of the output of the SPE before geometric transformation

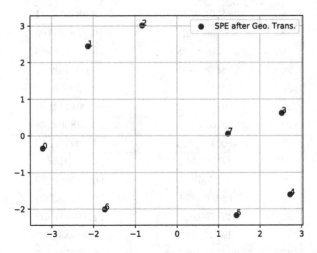

Fig. 11. A plot of the output of the SPE after geometric transformation

8   Computational Complexity

The computational complexity of the algorithm can be parametrized with respect to three different measures: (a) number of rounds, which is 1 in our case; (2) query complexity, which is the number of distance queries posed to the adversary and is a function of the number of rounds; (c) the time complexity of the algorithm.

The query complexity is the number of edges in the initial chordal graph. We have tried to make it as sparse as possible. It is always a tree on n vertices, with a few more edges added, to meet the requirements of the distance matrix completion algorithm. Thus the query complexity is $O(n)$. The time complexity of

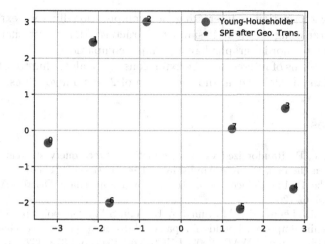

Fig. 12. A plot of the output of the YH vs SPE

the algorithm is dominated by the number of times we have to perform distance matrix completion of a clique. This is $O(n^2 f(n))$, as we go from a chordal graph which is nearly a tree to a complete (chordal) graph. This explains the n^2 term. The factor $f(n)$ is the complexity of the distance matrix completion algorithm. A loose upper bound is $O(n^3)$. Thus the time complexity of our algorithm is in $O(n^5)$.

9   Conclusion

In this paper we have proposed a 1-round algorithm for point placement in the plane in an adversarial setting, taking advantage of an existing infrastructure for completing partial distance matrices whose distance graphs are chordal. The locations of the points in the plane are recovered from the complete distance matrix, as outlined in Sects. 6 and 7.

Much more work remains to be done. The most interesting open question is this: for what kind of chordal graphs do we have a unique distance matrix completion? Bakonyi and Johnson [11] proved the following:

Theorem 7. *Let R be a partial distance matrix in R^k, the graph $G = (V, E)$ of whose specified entries is chordal and let S be the set of all minimal vertex separators of G. Then R admits a unique completion to a distance matrix if and only if*

$$\begin{pmatrix} 0 & e^T \\ e & R(S) \end{pmatrix}$$

has rank $k + 2$ for any $S \in \mathcal{S}$.

The characterization is interesting but computationally very expensive. It would be interesting to know if simpler characterizations exist that could be used to generate chordal graphs having unique completions.

Other problems of interest are the extensions of the algorithm to other classes of graphs than chordal graphs and the design of 2-round algorithms.

References

1. Damaschke, P.: Randomized vs. deterministic distance query strategies for point location on the line. Discrete Appl. Math. **154**(3), 478–484 (2006)
2. Damaschke, P.: Point placement on the line by distance data. Discrete Appl. Math. **127**(1), 53–62 (2003)
3. Chin, F.Y.L., Leung, H.C.M., Sung, W.K., Yiu, S.M.: The point placement problem on a line-improved bounds for pairwise distance queries. In: Giancarlo, R., Hannenhalli, S. (eds.) WABI 2007. LNCS, vol. 4645, pp. 372–382. Springer, Heidelberg (2007). https://doi.org/10.1007/978-3-540-74126-8_35
4. Alam, M.S., Mukhopadhyay, A.: More on generalized jewels and the point placement problem. J. Graph Algorithms Appl. **18**(1), 133–173 (2014)
5. Alam, M.S., Mukhopadhyay, A.: Three paths to point placement. In: Ganguly, S., Krishnamurti, R. (eds.) CALDAM 2015. LNCS, vol. 8959, pp. 33–44. Springer, Cham (2015). https://doi.org/10.1007/978-3-319-14974-5_4
6. Aspnes, J., et al.: A theory of network localization. IEEE Trans. Mobile Comput. **5**(12), 1663–1678 (2006)
7. Biswas, P., Liang, T., Wang, T., Ye, Y.: Semidefinite programming based algorithms for sensor network localization. TOSN **2**(2), 188–220 (2006)
8. Schoenberg, I.J.: Remarks to Maurice Fretchet's article "sur la definition axiomatique d'une classe d'espace distancis vectoriellement applicable sur l'espace de Hilbert". Ann. Math. **36**(3), 724–731 (1935)
9. Young, G., Householder, A.S.: Discussion of a set of points in terms of their mutual distances. Psychometrika **3**(1), 19–22 (1938)
10. Mukhopadhyay, A., Sarker, P.K., Kannan, K.K.V.: Point placement algorithms: an experimental study. Int. J. Exp. Algorithms **6**(1), 1–13 (2016)
11. Bakonyi, M., Johnson, C.R.: The euclidian distance matrix completion problem. SIAM J. Matrix Anal. Appl. **16**(2), 646–654 (1995)
12. Agrafiotis, D.K.: Stochastic proximity embedding. J. Comput. Chem. **24**(10), 1215–1221 (2003)
13. Rose, D.J., Tarjan, R.E., Lueker, G.S.: Algorithmic aspects of vertex elimination on graphs. SIAM J. Comput. **5**(2), 266–283 (1976)
14. Grone, R., Johnson, C.R., Sa, E.M., Wolkowicz, H.: Positive definite completions of partial hermitian matrices. Linear Algebra Appl. **58**, 109–124 (1984)
15. Bron, C., Kerbosch, J.: Algorithm 457: finding all cliques of an undirected graph. Commun. ACM **16**(9), 575–577 (1973)
16. Ellis, R., Lay, D.: Rank-preserving extensions of band matrices. Linear Multilinear Algebra **26**, 147–179 (1990)
17. Kabsch, W.: A solution for the best rotation to relate two sets of vectors. Acta Crystallogr. A **32**(5), 922–923 (1976)
18. Markenzon, L., Vernet, O., Araujo, L.H.: Two methods for the generation of chordal graphs. Ann. OR **157**(1), 47–60 (2008)

Neighbourhood Graphs and Locally Minimal Triangulations

Ivana Kolingerová[1,2]([⊠]), Tomáš Vomáčka[1], Martin Maňák[1,2],
and Andrej Ferko[3]

[1] Department of Computer Science, Faculty of Applied Sciences, University of West Bohemia, Pilsen, Czech Republic
{kolinger,tvomacka}@kiv.zcu.cz
[2] New Technologies for the Information Society, Faculty of Applied Sciences, University of West Bohemia, Pilsen, Czech Republic
manak@ntis.zcu.cz
[3] Department of Algebra, Geometry and Didactics of Mathematics, Faculty of Mathematics, Physics and Informatics, Comenius University, Bratislava, Slovakia
Andrej.Ferko@fmph.uniba.sk

Abstract. Neighbourhood (or proximity) graphs, such as nearest neighbour graph, closest pairs, relative neighbourhood graph and k-nearest neighbour graph are useful tools in many tasks inspecting mutual relations, similarity and closeness of objects. Some of neighbourhood graphs are subsets of Delaunay triangulation (DT) and this relation can be used for efficient computation of these graphs. This paper concentrates on relation of neighbourhood graphs to the locally minimal triangulation (LMT) and shows that, although generally these graphs are not LMT subgraphs, in most cases LMT contains all or many edges of these graphs. This fact can also be used for the neighbourhood graphs computation, namely in kinetic problems, because LMT computation is easier.

Keywords: Nearest neighbour graph · K-nearest neighbour graph
Locally minimal triangulation · Delaunay triangulation
Kinetic problem

1 Introduction

Neighbourhood (proximity) graphs rank among the most useful concepts in computational geometry and computational morphology, with numerous applications such as collision detection, image and document classification. The best known members of this graph family are the nearest neighbour graph and the Euclidean minimum spanning tree, however, many other exist, such as the Gabriel graph, the closest pairs (double nearest neighbour graph), the relative neighbourhood graph and the k-nearest neighbour graph. As these graphs contain edges connecting 'near' points, it is not surprising that all or some edges of these graphs

© Springer-Verlag GmbH Germany, part of Springer Nature 2018
M. L. Gavrilova and C. J. K. Tan (Eds.): Trans. on Comput. Sci. XXXIII,
LNCS 10990, pp. 115–127, 2018.
https://doi.org/10.1007/978-3-662-58039-4_7

can be found in Delaunay triangulation. This relation brings an easy way to compute all or some of the graph edges.

However, not all these graphs are subsets of Delaunay triangulation; so it is useful to have another source of the proximity graph edges. Locally minimal triangulation, although it is provably not a superset of the neighbourhood graphs, may provide a substantial amount of their edges. This fact may bring time savings in the computation of these graphs as this triangulation is easy to compute. There is so far no research devoted to the relation of the neighbourhood graphs to locally minimal triangulation. This research direction has seemed useless as locally minimal triangulation is rarely used in real applications – although its definition is simpler than that of Delaunay triangulation, the difference is not so substantial to justify the replacement of the Delaunay triangulation, with all its known and proved qualities and efficient algorithms to compute it.

The situation is different for kinetic data where the points or vertices move. The importance of a simpler condition for the edges or triangles to be tested in computation grows when the triangulation is to be recomputed again and again which is the case for kinetic points. This is a good reason to inspect whether there is a relation between the locally minimal triangulation and the neighbourhood graphs as, if so, the computation of the graph edges could be simplified.

This paper shows that although generally the neighbour graphs are not subgraphs of the locally minimal triangulation, many neighbour-graphs edges can be found in this triangulation and thus the replacement of the Delaunay triangulation in kinetic problems, such as a collision application, is possible.

Section 2 contains preliminaries for the static data, Sect. 3 concentrates on kinetic triangulations. Section 4 explains the relation between neighbourhood graphs and LMT. Section 5 presents experiments and results, Sect. 6 concludes the paper.

2   Preliminaries for Static Data

Let us survey definitions of the geometrical graphs used in the paper.

Definition 1 (Triangulation). *The triangulation $T(P)$ of a set of points P in the plane, $P = \{p_0, \ldots, p_{n-1}\}$, $n > 2$, is a maximum set of edges E such that*

- *edges from E intersect only in the points from P and*
- *edges from E subdivide the convex hull of P into triangles.*

The Delaunay triangulation definition is oriented rather to triangles than to edges:

Definition 2 (Delaunay triangulation). *The triangulation $DT(P)$ of a set of points P in the plane, $P = \{p_0, p_1, \ldots, p_{n-1}\}$, $n > 2$, is a Delaunay triangulation of P if and only if the circumcircle of any triangle of $DT(P)$ does not contain a point of P in its interior.*

It is useful to have also a condition of emptiness of the circumcircle formed for the edges of DT as for the purpose of neighbourhood graph edges search it is more proper to see the triangulations as graphs than as sets of triangles. Thus we include also the edge emptiness property:

Property 1 (Edge emptiness). An edge is in the Delaunay triangulation if and only if the edge has an empty circumcircle [7].

Delaunay triangulation has several well-known and important subgraphs – Euclidean minimum spanning tree (EMST), relative neighbourhood graph (RNG), Gabriel graph (GG), with EMST $\subseteq$ RNG $\subseteq$ GG $\subseteq$ DT [16]. These graphs are useful, first of all, in networks optimization and pattern recognition – they help to find structure among the points. Let us define those of them needed in the paper.

Definition 3 (Relative Neighbourhood Graph - RNG). *Let P be a set of points in the plane, $P = \{p_0, p_1, \ldots, p_{n-1}\}$, $n > 1$. The relative neighbourhood graph RNG is an undirected graph with P being its vertex set when for each edge between p_i and p_j there is no point p_k such that p_k is closer to both p_i and p_j than they are to each other: $d(p_i, p_j) \leq max\{d(p_i, p_k), d(p_j, p_k)\}$ [19].*

If we introduce a direction of the edges, we can define another subgraph of DT, a directed subgraph, called the nearest neighbour graph [15]:

Definition 4 (Nearest neighbour graph – $\overline{NNG}$.) *Let P be a set of points in the plane, $P = \{p_0, p_1, \ldots, p_{n-1}\}$, $n > 1$. $\overline{NNG}$ is a directed graph with P being its vertex set and with a directed edge from p_i to p_j whenever p_j is the nearest neighbour of p_i (i.e., the distance from p_i to p_j is no larger than from p_i to any other point from P) [16].*

$\overline{NNG}$ is sometimes replaced by an undirected graph, let us denote it NNG. There is a relation between NNG and EMST as follows: NNG $\subseteq$ EMST which implies also the relationship NNG $\subseteq$ DT. In general NNG is disconnected.

Definition 5 (Closest Pairs – CP/Double Nearest Neighbour Graph). *Let P be a set of points in the plane, $P = \{p_0, p_1, \ldots, p_{n-1}\}$, $n > 1$. CP is an undirected graph with P being its vertex set and with an edge $p_i p_j$ whenever p_j is the nearest neighbour of p_i and p_i is the nearest neighbour of p_j [20].*

Let us note that CP is usually disconnected, with the exception of singularities. CP $\subseteq$ NNG and therefore also CP $\subseteq$ DT.

An important member of the neighbourhood family is also the k-Nearest Neighbour Graph (k-NNG), defined as follows.

Definition 6 (k-Nearest Neighbour Graph - k-NNG). *Let P be a set of points in the plane, $P = \{p_0, p_1, \ldots, p_{n-1}\}$, $n > 1$. k-NNG is a directed graph with P being its vertex set and with k directed edges from p_i to its k nearest neighbours from P, for all points p_i.*

The k-NNG edges can be found in the DT of order k in $O(k^2 \, n \, log \, n)$ time. Some (but not all) of these edges are also present in DT [14].

Now let us proceed to other triangulations used in this paper:

Definition 7 (Locally minimal triangulation – LMT). *The triangulation LMT(P) of a set of points P in the plane, $P = \{p_0, p_1, \ldots, p_{n-1}\}$, $n > 2$, is a locally minimal triangulation of P if and only if every edge $p_i p_j$ shared by two triangles $p_i p_j p_k$, $p_i p_j p_l$ forming a convex quadrilateral is not longer than the diagonal $p_k p_l$ [9].*

It should be pointed out that LMT is not unique on the given point set P - more triangulations with locally minimal edges can be constructed. LMT is a well-known term in relation to the minimum weight triangulation. Usability of some of its edges to the minimum weight triangulation construction have been examined [1,4,5,8,9].

Definition 8 (Greedy triangulation – GT). *The triangulation GT(P) of a set of points P in the plane, $P = \{p_0, p_1, \ldots, p_{n-1}\}$, $n > 2$, is a greedy triangulation of P if and only if it consists of the shortest possible mutually non-intersecting edges.*

Existing research results and namely practical experiments show that different types of triangulations are for general data sets not so much mutually different as could be expected. Cho in [6] shows that at least 40% of DT edges are common with the edges of GT for points uniformly distributed in a region. According to his experiments, even about 90% of edges are common between the two triangulations. Kim et. al. in [13] found experimentally that more than 90% of the edges of a triangulated terrain model belong to DT and they utilize this fact in a compression.

Many algorithms exist to compute the DT triangulation - local improvement by edge flipping, divide & conquer, incremental insertion, incremental construction, sweeping and high-dimensional embedding. [18] provided a comparison of sequential DT algorithms; although nowadays a bit dated, still this study may provide useful general ideas about fundamental algorithmic strategies utilized to construct DT. [17] provides a set of benchmarks to check the correctness of the DT implementations.

LMT algorithms have not been so thoroughly inspected, however, either the incremental insertion or the local improvement by edge flipping are the easiest choice as the algorithm is nearly the same as for DT, the only difference being a different test on the validity of a triangulation edge. The non-uniqueness of LMT does not bring any problems - the algorithms converge to one of the triangulations satisfying the LMT definition.

3   Preliminaries for Kinetic Data

As shown in [2,12], we can say that spatial subdivision data structures are defined by functions of the input data called predicates which determine the resulting topology of the data structure depending on the input data. Each geometric structure constructed over a finite set of primitives may be proved valid by checking a finite number of predicates of these primitives. These checks are called certificates. In DT, the certificates are represented by the incircle test function which is a determinant formulation of the empty circumcircle test (1).

$$p_{DT} = det \begin{bmatrix} x_i & y_i & x_i^2 + y_i^2 & 1 \\ x_j & y_j & x_j^2 + y_j^2 & 1 \\ x_k & y_k & x_k^2 + y_k^2 & 1 \\ x_l & y_l & x_l^2 + y_l^2 & 1 \end{bmatrix} \tag{1}$$

In LMT the certificates are comparisons of lengths of the two possible edges - diagonals of the convex quadrilateral formed by two neighbouring triangles (2).

$$p_{LMT} = (x_i - x_j)^2 + (y_i - y_j)^2 - (x_k - x_l)^2 - (y_k - y_l)^2 \tag{2}$$

where p_i, p_j, p_k and p_l are points from P.

As we are considering the kinetic counterparts of these data structures, it is necessary to note that the coordinates of the source data are in fact functions of time. They are most commonly in the form of polynomials, but they may be any functions as long as it is possible to solve them to find their roots with a necessary precision.

In order to maintain a kinetic data structure, one has to maintain a priority queue of the currently active predicates sorted by the increasing value of the nearest next root (greater than the current time value). Each time the value of current time reaches the value determined by the first certificate function in the queue, at least one of the certificates becomes invalid and the topology of the kinetic data structure needs to be changed. This change depends on the type of the data structure - in the case of kinetic Delaunay triangulation (KDT) and kinetic locally minimal triangulation (KLMT) they are represented by an edge swap in the triangle pair associated with the predicate (and certificate). Also, usually there are new certificates that need to be placed into the queue. Swap conditions for moving circles and line segments can be found in [10] and for spheres in [11].

From the facts stated above it is obvious that in order to maintain a kinetic data structure one has to solve a large number of algebraic equations. Let us now have a set of points $P = \{p_0, p_1, \ldots, p_{n-1}\}$, $n > 2$, KDT(P) and KLMT(P). Let also $p_i = (x_i(t), y_i(t))$, where $x_i(t)$ and $y_i(t)$ are polynomial functions of degree $R \geq 1$.

Lemma 1. *The computation of certificates for KDT(P) is at least as computationally complex as for KLMT(P) constructed over the same set of kinetic points moving along polynomial trajectories of degree up to R.*

Proof. If the two data structures are constructed over a set of static points ($R = 0$, thus voiding their kinetic property), their kinetic maintenance is no longer necessary which applies to both of the structures. In every other case we can see from (1) that in order to compute the predicates for KDT(P), we need to search periodically for the roots of polynomials of the degree up to $4R$ and from (2) we can see that in case of KLMT(P) we only need to solve polynomials of degree up to $2R$.

From the practical point of view, it may be worth mentioning that the polynomials $x_i(t)$ and $y_i(t)$ are often of degree one (meaning that the points move along linear trajectories) which results in solving quadratic equations for KLMT and fourth-degree polynomials for KDT. Note that solving fourth-degree polynomials usually cannot be done analytically which further increases the overall complexity of maintaining the KDT, especially when compared to KLMT.

Having all necessary information, we can proceed to the relation between neighbourhood graphs and LMT, the core topic of this paper.

4   Relation of Neighbourhood Graphs and LMT

The expectation that neighbourhood graphs edges are not only subsets of DT but also of LMT is supported by the fact that LMT contains "short edges". Unfortunately, the situation is not so nice. It is easy to show that, generally, NNG is not contained in LMT.

Lemma 2. *Let us have a LMT on the set of points P. Then NNG of the given set of points can contain edges which do not belong to LMT.*

Proof. To prove the lemma it is enough to find a data set P, which will produce an edge in NNG that will not belong to LMT. Such a data set can be found even on four points: $P = \{p_0, p_1, p_2, p_3\}$. If the line segments p_0p_1 and p_2p_3 intersect in exactly one point $\notin P$ and if $|p_2p_3| < |p_0p_1| < min(|p_0p_2|, |p_0p_3|)$ then the points of P constitute a convex quadrilateral with diagonals p_0p_1 and p_2p_3, the edge from p_0 to p_1 is in NNG but not in LMT since p_2p_3 is shorter than p_0p_1.

Figure 1 illustrates the proof of the lemma. Points p_0 and p_1 can be chosen arbitrarily, then the point p_2 is chosen from the first highlighted region and after that, the point p_3 is chosen from the second highlighted region. The circles have centres in p_0, p_1, p_2 and they all have the same radius $|p_0p_1|$.

From this example it is evident that NNG is generally not included in LMT. It leads to suspicion that also CP and RNG are generally not subsets of LMT. But due to the fact that triangulations used in real applications have a considerable common subset of edges, it can be expected that many edges of neighbourhood graphs can be produced from LMT. As concerns k-NNG, for $k > 1$ not all its edges are contained in DT, so similar behaviour for LMT can be expected.

In real applications, e.g., for collision tests in virtual reality, where savings in computation can be substantial, some small omissions in neighbourhood graphs can be tolerated in favour of increased efficacy of computation. Thus in the next section we will concentrate on the expected behaviour of the DT, LMT and neighbourhood graphs in experiments for various data. Our effort is to show that most of the neighbourhood graphs edges are mostly present in LMT in spite of the fact that one-hundred-percent-inclusion cannot be guaranteed.

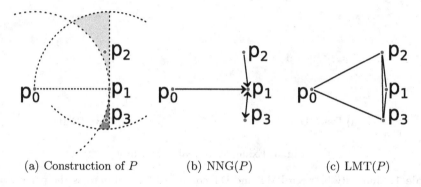

(a) Construction of P (b) NNG(P)   (c) LMT(P)

Fig. 1. A set $P = \{p_0, p_1, p_2, p_3\}$ of points for which NNG(P) $\not\subseteq$ LMT(P)

5   Experiments and Results

Testing data sets were both artificially generated data sets with randomly generated points distributed uniformly, in clusters, with Gaussian distribution and on an arc, see Fig. 2, and real terrain data (kindly provided by Bayer [3]). Let us remind that DT and LMT methods use only planar part of the terrain data. The data sets did not contain any duplicated points. The cluster data are represented by 10 uniformly generated points used as a centre of a Gaussian distribution. The artificial data sets were from 100 up to 100000 points, the real data from about 11000 up to about 150000. Examples of real data sets are depicted in Fig. 3. As there can be more LMT on one data set, each test was repeated for several LMT versions but the differences in results were negligible.

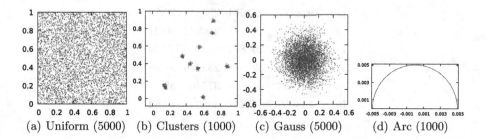

(a) Uniform (5000)   (b) Clusters (1000)   (c) Gauss (5000)   (d) Arc (1000)

Fig. 2. Examples of artificial data sets used in experiments

5.1   Similarity Between DT and LMT

Similarity S of two triangulations is given by (3).

$$S = \frac{ne_{LD}}{ne_L}100 \; [\%] \tag{3}$$

(a) Data ID 9   (b) Data ID 10

Fig. 3. Examples of real data sets

Table 1. Similarity between LMT and DT computed for real data as the percentage $S\%$ of LMT edges that are also present in DT. Columns ne_L contain the total number of LMT edges, ne_{LD} the number of LMT edges in DT.

Data ID	Vertices nv	Edges ne_L	ne_{LD}	$S\%$
1	11495	34457	30835	89.49
2	18552	55623	50174	90.20
3	21407	63502	57609	90.72
4	25392	74194	66486	89.61
5	26207	78588	75216	95.71
6	30264	86372	77524	89.76
7	49391	148146	132322	89.32
8	74110	221423	196823	88.89
9	78659	235946	210409	89.18
10	79903	239678	215309	89.83
11	92845	275664	250548	90.89
12	98616	295807	266949	90.24
13	100632	301863	272425	90.25
14	110369	326589	289296	88.58
15	141045	423083	375461	88.74
16	148559	445641	398941	89.52

where ne_L is the total number of edges of LMT, ne_{LD} the number of common edges of LMT and DT.

Table 1 shows results of DT and LMT comparison for real data.

It can be seen from Table 1 that similarity of both types of triangulations is relatively high for real data, DT and LMT have about 90% edges in common. No dependence of results on the size of data set was detected. These results are in correspondence with the previous observation in [6,13].

Table 2 shows similarity for artificially generated data. Gauss data indicate a bit lower similarity but still about 85%. Uniform data provided similar percentage, but the decrease of S with growing triangulation size. Cluster data presented very low similarity for two largest data sets. Arc data have only about 59% similarity.

Although the cases with high similarity, corresponding to the previous results from [6,13], were prevailing in the tests, during experiments we came also across the data sets which had a substantially lower similarity of LMT and DT, such as cluster data sets in the rows 6 and 7 of Table 2. These were the cases of data sets which contained many configurations with concentric points. Let us consider a small data set of several points sitting on a circle empty from other points, then DT of this point group is ambiguous – there are more than one DT. The simplest example is a group of four points in a square position, for which any of both square diagonals (but only one of them) is present in DT. In such a case even two DTs of the same set would have lower similarity than 100%. If many such cases are present in the testing data sets, similarity of two triangulations is decreased.

Table 2. Similarity between LMT and DT computed for sets of random points (various distributions) as the percentage $S\%$ of LMT edges that are also present in DT. Columns ne_L contain the total number of LMT edges, ne_{LD} the number of LMT edges in DT.

Vertices	Gauss			Uniform			Clusters			Arc		
nv	ne_L	ne_{LD}	$S\%$	ne_L	ne_{LD}	$S\%$	ne_L	ne_{LD}	$S\%$	ne_L	ne_{LD}	$S\%$
100	285	252	88	286	262	92	280	241	86	102	64	63
500	1486	1294	87	1479	1307	88	1481	1305	88	243	144	59
1000	2984	2633	88	2982	2641	89	2987	2635	88	335	183	55
5000	14980	13191	88	14966	13063	87	14980	13185	88	1019	361	35
10000	29965	25738	86	29967	26281	88	29978	26386	88	25594	19188	75
50000	149620	118979	80	149870	127395	85	149952	46377	31	47876	31903	67
100000	298596	231843	78	299613	228589	76	299917	51960	17	63695	38101	60

5.2 NNG Edges Inclusion in LMT

Next we tested how many NNG edges (computed from DT) can be found in LMT, see Table 3. The tests were done on condition that each vertex has only one nearest neighbour – i.e., for each vertex, only one outgoing edge was examined. Such a condition would be improper for degenerated cases, such as the points sitting on a regular grid but such data are usually not subject to DT as they can be triangulated directly.

It can be seen that all or nearly all NNG edges can be found in LMT, even for the triangulations which showed a lower mutual similarity in the previous group of tests.

Table 3. The percentage of NNG edges (*ne*) found also in LMT for real data sets and for various artificial data sets

Real data sets						Artificial data sets				
Data	Vertices	Edges	Data	Vertices	Edges	Vertices	Gauss	Uniform	Clusters	Arc
ID	*nv*	*ne*[%]	ID	*nv*	*ne*[%]	*nv*	*ne* [%]	*ne* [%]	*ne* [%]	*ne* [%]
1	11495	99.93	9	78659	99.95	100	100	100	100	100
2	18552	99.91	10	79903	92.94	500	99.80	100	99.60	100
3	21407	98.42	11	92845	98.92	1000	99.90	100	100	100
4	25392	96.79	12	98616	99.94	5000	99.94	100	99.98	100
5	26207	99.99	13	100632	99.95	10000	99.99	100	99.97	100
6	30264	93.37	14	110369	98.16	50000	99.98	100	99.89	100
7	49391	99.90	15	141045	99.96	100000	99.98	100	99.79	100
8	74110	99.43	16	148559	99.96					

5.3   CP and RNG Inclusion in LMT

Next we inspected how many CP edges (computed from DT) can be found in LMT. Results for CP are in Table 4 and results for RNG in Table 5.

Table 4. The percentage of CP edges (*ne*) found also in LMT for real data sets and for various artificial random data sets

Real data sets						Artificial data sets				
Data	Vertices	Edges	Data	Vertices	Edges	Vertices	Gauss	Uniform	Clusters	Arc
ID	*nv*	*ne*[%]	ID	*nv*	*ne*[%]	*nv*	*ne*[%]	*ne* [%]	*ne* [%]	*ne* [%]
1	11495	100	9	78659	100	100	100	100	100	100
2	18552	100	10	79903	100	500	100	100	100	100
3	21407	98.98	11	92845	99.06	1000	100	100	100	99.70
4	25392	97.90	12	98616	100	5000	100	100	100	96.85
5	26207	99.40	13	100632	100	10000	99.97	100	100	97.31
6	30264	95.40	14	110369	98.87	50000	100	99.96	100	99.99
7	49391	100	15	141045	100	100000	99.97	99.98	100	99.76
8	74110	99.69	16	148559	100					

These results show that LMT can be a very good source of CP and RNG edges as the amount of unfound graph edges is usually below one per cent. However, we should stress out that LMT can also provide some false positives for CP and RNG graphs - some edges which are detected as neighbourhood graph edges, although they do not belong there. The amount of these false positives is very small, under 0.1 per cent with the exception of arc data where the amount is a bit higher, up to 0.8 per cent.

Table 5. The percentage of RNG edges (*ne*) found also in LMT for real data sets and for various artificial random data sets

Real data sets						Artificial data sets				
Data	Vertices	Edges	Data	Vertices	Edges	Vertices	Gauss	Uniform	Clusters	Arc
ID	*nv*	*ne*[%]	ID	*nv*	*ne*[%]	*nv*	*ne* [%]	*ne* [%]	*ne* [%]	*ne* [%]
1	11495	99.43	9	78659	99.41	100	100	100	99.05	100
2	18552	99.43	10	79903	99.37	500	99.04	99.35	99.31	100
3	21407	98.46	11	92845	98.61	1000	99.52	99.60	100	100
4	25392	97.35	12	98616	99.45	5000	99.31	99.29	99.35	100
5	26207	99.01	13	100632	99.45	10000	99.48	99.36	99.32	100
6	30264	94.96	14	110369	98.19	50000	99.27	99.35	99.36	100
7	49391	99.30	15	141045	99.21	100000	99.21	99.43	99.36	100
8	74110	98.97	16	148559	99.37					

5.4   K-NNG Inclusion in LMT

As k-NNG edges for $k > 1$ may not be contained in DT, the tests were done in a different way. First k-NNG graphs for testing datasets were computed for $k = 2$, $k = 3$ by brute force, then these edges were sought in DT and LMT. Due to higher time demands of k-NNG computations, smaller datasets than in previous experiments were tested. Results are a bit surprising. LMT was found a bit better source of 2- and 3-neighbours than DT, see Tables 6 and 7.

Table 6. The percentage of 2-NNG and 3-NNG edges found in DT and LMT for real data sets

Data	Vertices	2-NNG edges		3-NNG edges	
ID	*nv*	in DT	/LMT	in DT	/LMT
1	11495	94.67	95.99	76.85	79.98
2	18552	95.05	96.28	78.55	81.27
3	21407	92.54	94.02	77.22	80.16

Table 7. The percentage of 2-NNG and 3-NNG edges found in DT and LMT for uniform and cluster data sets

Uniform					Clusters				
Vertices	2-NNG edges		3-NNG edges		Vertices	2-NNG edges		3-NNG edges	
nv	in DT	/LMT	in DT	/LMT	*nv*	in DT	/LMT	in DT	/LMT
100	96.00	97.00	80.00	82.00	100	91.00	94.00	66.00	68.00
500	91.80	93.00	73.00	73.50	500	90.80	94.60	68.40	72.80
1000	93.20	95.10	73.60	77.50	1000	92.00	94.70	70.80	74.60
5000	92.22	95.76	72.02	78.04	5000	92.67	95.42	72.66	76.72
10000	92.70	95.35	73.00	78.35	10000	92.65	95.07	73.11	77.83

6   Conclusion

From these experiments a conclusion can be drawn that for practical applications, LMT can serve as a good source of neighbourhood graphs edges, especially for k-KNN graphs. It may be interesting namely for applications with kinetic data, such as collision detection or collision avoidance, where kinetic triangulations are needed, as even the small savings achieved by a simpler predicate/certificate computation in the kinetic LMT can be important due to a high number of their recomputations.

In future work it would be interesting and useful to examine the 3D versions of the triangulations and their relation to neighbourhood graphs where even higher impact on collision detection tests can be expected.

Acknowledgements. This work was supported by the Czech Science Foundation, the project number 17-07690S, and by the Ministry of Education, Youth and Sports of the Czech Republic, project number LO1506 (PUNTIS). We would like to thank to T. Bayer from the Charles University in Prague, Czech Republic for supplying us the real terrain data for the experiments.

References

1. Aichholzer, O., et al.: Triangulations intersect nicely. Discrete Comput. Geom. **16**(4), 339–359 (1996)
2. Basch, J., Guibas, L.J., Hershberger, J.: Data structures for mobile data. J. Algorithms **31**(1), 1–28 (1999)
3. Bayer, T.: Department of Applied Geoinformatics and Cartography, Faculty of Science, Charles University, Prague, Czech Republic. https://web.natur.cuni.cz/~bayertom. Accessed 15 May 2017
4. Beirouti, R., Snoeyink, J.: Implementations of the LMT heuristic for minimum weight triangulation. In: Proceedings of the Fourteenth Annual Symposium on Computational Geometry, SCG 1998, pp. 96–105. ACM, New York (1998)
5. Bose, P., Devroye, L., Evans, W.: Diamonds are not a minimum weight triangulation's best friend. Int. J. Comput. Geom. Appl. **12**(06), 445–453 (2002)
6. Cho, H.G.: On the expected number of common edges in Delaunay and greedy triangulation. J. WSCG **5**(1–3), 50–59 (1997)
7. Dey, T.K.: Curve and Surface Reconstruction: Algorithms with Mathematical Analysis (Cambridge Monographs on Applied and Computational Mathematics). Cambridge University Press, New York (2006)
8. Dickerson, M.T., Keil, J.M., Montague, M.H.: A large subgraph of the minimum weight triangulation. Discrete Comput. Geom. **18**(3), 289–304 (1997)
9. Dickerson, M.T., Montague, M.H.: A (usually?) connected subgraph of the minimum weight triangulation. In: Proceedings of the Twelfth Annual Symposium on Computational Geometry, SCG 1996, pp. 204–213. ACM, New York (1996)
10. Gavrilova, M., Rokne, J.: Swap conditions for dynamic Voronoi diagrams for circles and line segments. Comput. Aided Geom. Des. **16**(2), 89–106 (1999)
11. Gavrilova, M., Rokne, J.: Updating the topology of the dynamic Voronoi diagram for spheres in Euclidean d-dimensional space. Comput. Aided Geom. Des. **20**(4), 231–242 (2003)

12. Guibas, L., Russel, D.: An empirical comparison of techniques for updating Delaunay triangulations. In: Proceedings of the Twentieth Annual Symposium on Computational Geometry, SCG 2004, pp. 170–179. ACM, New York (2004)
13. Kim, Y.S., Park, D.G., Jung, H.Y., Cho, H.G., Dong, J.J., Ku, K.J.: An improved TIN compression using Delaunay triangulation. In: Proceedings of Seventh Pacific Conference on Computer Graphics and Applications (Cat. No.PR00293), pp. 118–125 (1999)
14. Maus, A., Drange, J.M.: All closest neighbors are proper delaunay edges generalized, and its application to parallel algorithms. In: Proceedings of Norwegian informatikkonferanse, pp. 1–12 (2010)
15. Okabe, A., Boots, B., Sugihara, K., Chiu, S.N.: Spatial tessellations: concepts and applications of Voronoi diagrams. Probability and Statistics, 2nd edn. Wiley, NYC (2000)
16. Preparata, F.P., Shamos, M.: Computational Geometry: An Introduction. Springer, New York (1985). https://doi.org/10.1007/978-1-4612-1098-6
17. Spelič, D., Novak, F., Žalik, B.: Delaunay triangulation benchmarks. J. Electr. Eng. **59**(1), 49–52 (2008)
18. Su, P., Drysdale, R.L.S.: A comparison of sequential Delaunay triangulation algorithms. Comput. Geom. **7**(5), 361–385 (1997)
19. Toussaint, G.T.: The relative neighbourhood graph of a finite planar set. Pattern Recogn. **12**(4), 261–268 (1980)
20. Veltkamp, R.C.: The γ-neighborhood graph. Comput. Geom. **1**(4), 227–246 (1992)

Author Index

Printed in the United States
By Bookmasters